Robert Boyle

Tracts Consisting of Observations About the Saltness of the Sea

an account of a statical hygroscope and its uses - together with an appendix about

the force of the air's moisture - a fragment about the natural and preternatural

state of bodies

Robert Boyle

Tracts Consisting of Observations About the Saltness of the Sea
an account of a statical hygroscope and its uses - together with an appendix about the force of the air's moisture - a fragment about the natural and preternatural state of bodies

ISBN/EAN: 9783337780623

Printed in Europe, USA, Canada, Australia, Japan

Cover: Foto ©berggeist007 / pixelio.de

More available books at **www.hansebooks.com**

Confisting of

OBSERVATIONS

About the

SALTNESS of the SEA:

An Account of a

STATICAL HYGROSCOPE

And its USES:

Together with an APPENDIX
about the

FORCE of the AIR'S MOISTURE:

A FRAGMENT about the

NATURAL and PRETERNATURAL
STATE of BODIES.

By the Honourable *ROBERT BOYLE.*

To all which is premis'd

A SCEPTICAL DIALOGUE

About the POSITIVE or PRIVATIVE
NATURE of COLD:

With fome Experiments of Mr. *BOYL'S* referr'd
to in that Difcourfe.

By a Member of the *ROYAL SOCIETY.*

London, Printed by *E. Flefher* for *R. Davis* Bookfeller
in *Oxford,* . M DC LXXIV.

AN

ADVERTISEMEMT

OF THE

PUBLISHER.

THE Reader of the following Dialogue may eafily conclude from the beginning of it; (where the occafion of the Conference is fet down) that

if

if the Author had been so minded, it might have long since come abroad. But though, as his backwardness to publish it kept it long lying by, first in His hands, and then in Mine; yet the Affinity it has with some of the ensuing Tracts of *Mr. Boyle,* and some other of his Papers, that he design'd (but was hinder'd) to have added to them, engag'd me to take the

liber-

liberty of publishing it and them all together : Which I had sooner done than now I do, if, by some Accidents, they had not been kept from appearing for many weeks after they were quite printed off.

A

OF THE

POSITIVE

OR

PRIVATIVE NATURE

OF

COLD.

A Sceptical Dialogue

Between

Carneades, Themiſtius, Eleutherius,
Philoponus.

SECT. I.

Eleu. **M**Ay one be allowed to ask *Carneades,* what Book it is he is reading with ſo much attention?

Carn. The Queſtion, *Eleutherius,* is very allowable, and as eaſily anſwer'd, by ſaying, that what I was reading, is our Friend Mr. *Boyle's* newly publiſhed *Hiſtory of Cold.*

The-

Themiſtius. Your readineſs, *Carneades*, to anſwer, encourages me alſo to ask you a Queſtion ;
which ſhall not be , as probably you expect it
ſhould, How you like this new Piece ? for I know
you would be too kind to the Author, not to tell me
that he has detected ſome Old Errours , and made
diſcovery of ſome New Truths ; but my Queſtion
ſhall be about what is my Wonder, as well as that
of divers others, who think it ſtrange that a Writer
that has deliver'd ſo many Effects and other *Phænomena of Cold*, ſhould omit to tell us ſo much as
whether he aſſerts it to be a *Poſitive* Quality, or a
bare *Privation* of Heat ; as, ſince *Cardan* (in his
Treatiſe *De Subtilitate*) ſome other Learned Men,
and eſpecially *Cartesius*, hath maintain'd.

Carneades. You will not wonder, if a Perſon that
you look upon, and I confeſs not injuriouſly, as
a Friend to Mr. *Boyle*, tell you, that this Author,
by the many Hiſtories he has preſented us, and by
his not ſeeming to dare to determine the Controverſie you have mention'd , ſhews, that he was
more ſollicitous to leſſen his ignorance , than to
pretend to knowledge : And upon the obſervation
I have made of his humour in general, I preſume
one principal reaſon of his ſilence may be, that he
has not yet compleated the trials he had deſign'd
about *Cold* ; and thinks, that in Abſtruſe Subjects,
ſuch as this is, 'tis not ſo convenient to deliver a
poſitive opinion of the Nature of it at the Beginning, as to reſerve it for the latter End, after the
Hiſtory of the *Phænomena* ; when the nature of
the thing inquir'd into may, as it were, ſpontáne-
ouſly Reſult from the Conſiderations ſuggeſted by
the precedent matters of Fact ſurvey'd together.

Eleu-

Eleutherius. If such a wariness were indeed the motive of your friends silence, I shall easily excuse it; and perhaps think too, that the like would not mis-become Naturalists on many other occasions. And yet I do not dislike *Themistius*'s question; for 'tis one thing to venture upon declaring the adæquate Nature of Cold, and another to determine, Whether it be a Positive, or a Privative Quality? The latter attempt importing a much less venture than the former.

- *Carneades.* I will not pretend to know the very Reasons that induc'd the Author silently to pass by this Controversie; but having been once present, when he had occasion to discourse of it, I then conjectur'd, that among his Experiments of *Cold*, that are not yet publish'd, there may be some uncommon ones, that may have suggested to him scruples, which oblig'd him to forbear declaring himself, till he had clear'd them, which those that are unacquainted with such Tryals, may probably have never thought of.

Themistius. If what you call a Controversie, were indeed worthy of that name, I should not unwillingly allow of your Friends silence; but the Opinion broach'd by *Cardan*, and adopted by Mr. *Des Cartes* and others, seems to me so devoid, not only of reason, but of all appearance of it, that me thinks one that has deliver'd such considerable Effects of Cold, as Mr. *Boyle* has done, may well ascribe to their cause, at least, a *Positive* Nature; and, without at all being guilty of boldness, reject an Opinion, that is not only barely an Errour, but an Extravagance, and perhaps a plain Absurdity.

Car-

Carneades. Possibly the Gentleman we are speaking of, may be wary and sceptical enough to reckon among difficult things, not onely the declaring the adæquate Nature of Cold, and the manner of its Operations ; but the demonstrating whether it be a *Positive* Quality or not. And though I will not take upon me to know his thoughts about that subject, which, perhaps, are grounded upon some of his peculiar Experiments and Notions ; yet, for discourse sake, I am content to debate with *Themistius*, Whether or no the Opinion he so severely censures, be not only erroneous, as, for ought appears, Mr. *Boyle* himself may be found to have thought it ; but also, as *Themistius* would have it, absurd.

Themistius. I readily accept of your offer ; for it cannot be an unpleasant entertainment to observe the arts whereby one that I know will not speak impertinently, will endeavour to make Reason elude the clearest Testimonies of Sense. And though I might press you with the concurrent authority of *Aristotle*, and all the Philosophers that have liv'd between his time, and those of that extravagant fellow *Cardan* ; yet I shall rather employ, to convince you, the authority and reasons of a grand Leader among your New Philosophers, who being a great broacher of Paradoxes, and having upon that score written Books expresly against *Aristotle*, was not like to have sided with him, unless the Evidence of Truth had, as it were, necessitated him to do so.

Carneades. I presume, you mean the Learn'd and Subtle *Gassendus*, whom I am glad you have pitch'd upon for your Causes Champion, not only because

in defending the common opinion, he waves the
common practice of troubling his Readers with a
multitude of Authorities, which to me, in such a
case as this, would signifie very little, and betakes
himself to arguments ; but because, being so modern and judicious a Writer, we may well suppose
him to have summ'd up and improv'd what can be
said in behalf of the cause he maintains. Upon
which account, I shall be excus'd from answering
impertinent Objections against the Opinion I defend, and from the trouble of ranging about among
other Authors for more weighty Arguments than
those, which the disproving of his will shew to be
unsatisfactory.

Themistius. I am glad you nam'd the Author I
meant, *Carneades* ; for I apprehended you had not
met with what he sayes upon this subject : because
I could scarce imagine, that an intelligent person,
after having read his arguments, will doubt of a
Truth he hath so clearly evinc'd by them. But since
I perceive you have seen what he has written, I
shall, without farther preamble, propose his Reasons to you, though not in the very same order
wherein he has couch'd them.

Eleutherius. But before you begin them, give me
leave to ask *Carneades* a short question, whose answer will, I suppose, conduce, if not be necessary,
to the clearing of the state of the Controversie betwixt you. For 'tis one thing to deny belief to the
receiv'd Opinion, that Cold is a Positive Quality,
and another thing to assert, that 'tis but a Privation
of Heat ; since, if *Carneades* does undertake the
latter of these two, he must bring positive Arguments to prove Cold to be but a negative thing.

Whereas, if he content himself to play a doubting part, it may suffice him, being in effect but a Defendant, to shew that the proofs brought to conclude Cold to be a Positive Quality, are not Cogent.

Carneades. I acknowledge your Question, *Eleutherius*, to be pertinent, and not unseasonable. And I presume, you will not be surpriz'd, that a Person accus'd of Scepticism answers it by declaring, that he undertakes not to demonstrate, that *Cold* must be a Privative or Negative Quality, and thinks it sufficient for his turn, to shew that the Arguments brought to evince it to be a Positive one, are not concluding. And, since you have already diverted *Themistius* from beginning so soon as he intended, 'twill not be amiss, that I continue that suspension a little longer, to prevent, what I know we both hate, Verbal Controversies; which yet may very easily spring from undetermin'd acceptions of Words as ambiguous as I have observ'd Heat (of which I now make Cold but a Privation) to be.

We may therefore consider, that the word *Heat*, being made use of to signifie, as well the operations of that quality upon other Bodies (as when the Heat of the Fire makes Water boyl, or that of the Sun melts Wax, and hardens Clay) as its operations upon the Sense of man, (as when a moderate degree of Heat is said to cause pleasure, and an excessive one to produce pain;) this Term, I say, as Mr. *Boyle* also has somewhere noted, may be employ'd sometimes in a more absolute and indefinite sense, and sometimes in a more confin'd and respective sense: In the latter of which, 'tis estimated

ted

ted by its Relation to the Organs of Feeling of
those men that judge of it.　Upon which account,
men are wont to esteem no body Hot, but such an
one, the agitation of whose small parts is brisk e-
nough to increase or surpass that of the particles of
the Organ that touches it.　For if that motion be
more Languid in the Object than in the Sentient,
the Body is reputed Cold; as may appear by this,
that if the same Person put one of his hands when
'tis hot, and the other when 'tis cold, into luke-warm
Water, that Liquor will feel cold to the warm hand,
and warm to the cold.

　Eleutherius. So that according to this Doctrine,
methinks, one may, for brevities sake, conveniently enough apply to your two-fold Notion of Heat,
those expressions which some School-men employ
about certain Qualities, of any of which they say,
that it may be either materially or formally consider'd.　And by Analogy to their Doctrine, since
Heat is a Tactile Quality, and as such, imports
primarily a relation to the Organ of Touching, that
relation, with what depends upon it, may pass for
that which is the *Formale* in the Quality called
Heat; and its Effects and Operations upon other
Bodies may supply us with a Notion of Heat, materially taken.

　Carneades. I do not alwayes quarrel, *Eleuthe-*
rius, with Terms borrow'd from the Schools, if they
be as much more short and expressive than others,
as they are more unusual, or even barbarous.　But
there is another Distinction of *Heat,* partly grounded upon that already propos'd; which, because it
may be of use in our future Discourse, will not be
unfit to be here intimated.　For we may consider,

　　　　　　that

that though, for the most part, a hot Body is taken
in the vulgar sense for that wherein the degree of
Heat is sensible to our Organs of Feeling; yet in
a looser sense, and which, for Distinctions sake,
we may call Philosophical, because concluded by
Reason, though not perceiv'd by sense, a Body may
be conceiv'd not to be destitute of Heat, even
when the degree of that Quality is not great enough
to be felt by the Touch; provided it can produce
in some degree those other Operations, which,
when more intense, are acknowledged to proceed
from manifest Heat. For elucidation of which, we
may alledge, That in very frosty, and yet clear, Weather, the Sun may be judg'd to warm the Air, when
it melts Snow, and thaws Ice; though, perhaps,
many men, especially of tender Constitutions, feel
in their Fingers and Toes much stiffness and more
pain, upon the account of Cold. To this I may
add the common Observation, if you grant the truth
of it, that Snow melts much sooner upon Land
newly turn'd up by the Plow, than, *cæteris paribus*,
in the neighbouring ground; which argues a
warmth in that newly expos'd earth : though according to the Touch it would questionless appear
Cold. But we may be furnish'd with a clearer and
more pregnant Instance, by but recalling to mind
what was just now mention'd of the warmth of tepid water, which was not to be felt by a hot
hand, but produc'd there a contrary sensation of
Cold. Which Instance I therefore scruple not to
repeat, because it affords an Experiment in favour of that premis'd Distinction, which, I think,
may also have this ground in Reason, that a considerable Heat is often requisite to be sensible to our
hands,

hands, *&c.* which are continually irrigated with the Circulating Blood that comes very warm out of the Heart, and enliven'd by Animal Spirits, plentifully fupply'd from the Brain.

If *Eleutherius* think fit to accommodate this Diftinction in the Vulgar and in the Philofophical fenfe to his *Heat,* formally and materially taken, I leave him to his liberty. And I fhall alfo leave it to you both, Gentlemen, to accommodate to Cold, *mutatis mutandis,* as they fpeak, what has been faid about the diftinctions of Heat; becaufe, I fear, *Themiftius* thinks himfelf to have been too long detain'd already from propofing his Arguments, which he may now begin to do affoon as he pleafes.

SECT. II.

Them. I Will then, with your permiffion, begin with that Argument of *Gaffendus,* which I am able to give you in his own words; becaufe upon the occafion of Mr. *Boyle*'s book, I made a Tranfcript of what he fayes to evince the *Pofitive* Nature of Cold; and having the Tranfcript yet about me, 'tis eafie for me to tell you, that 'tis this : *Ii funt frigoris effectus quales habere privatio, quæ actionis eft incapax, non poteft.*

Vid. *Gaffend. Phyficam. Sect.1.Lib. 6. Cap.6.*

This Argument, though he begins not with it, I choofe to make the firft, becaufe I think it of fuch weight, that, though it were the only one he could alledge, it would ferve his turn and mine, fince 'tis drawn from the Effects of Cold, which, though he

men-

mentions them but in few and general words, experience shews to be both so manifold and so considerable, that if *Carneades* imploy an hundred times as much time to answer the Argument they afford, as I have done to recite it, he will, I think, do no more than would be necessary, and perhaps not enough to be sufficient. For, *Cold* affects the Organs of Feeling, and sometimes causes great pain in them, condenses Air and Water, and breaks Bottles that are too well stopt, congregates both Homogeneous and Heterogeneous things, increases Hunger, checks fermentation in Liquors, produces Heat by *Antiperistasis*, in deep Cellars, Mines, *&c.* and yet freezes Men and Beasts to Death, dismantles whole Woods and Forrests of their Leaves, and does (I know not how many) other Feats ; among which, it is not the least admirable, though one of the most common, that it turns the fluid and yielding Waters of Rivers and Lakes, and sometimes of part of the Sea it self, not too far from the shoar, into firm and solid Ice, which is often in Northern Climates strong enough, not only to be travell'd upon by Merchants with their Carriages, but to be fought upon by whole Armies with their trains of Artillery. From which, and other Instances, it is manifest, that Effects so numerous and great, cannot proceed from a meer privation, or any negative thing, but require a considerable, and therefore sure a Positive, Quality to produce them.

Carneades. This Objection , *Themistius*, is , I confess, a considerable one, and of more weight than any of the rest, if not than all of them put together : But, as I think it very worthy to be answer'd,

fwer'd, fo I think it very poffible to be well anfwer'd; and to give you my reafons for my fo thinking, I fhall diftinctly confider in the Argument
the two particulars which it feems to confift of.

· And firft we are told, that if Cold be but a Privation, it cannot be the object of fenfe. To clear
this difficulty, which, I know, you will think it very
hard, if at all poffible to do, I muft beg your leave
to obferve fomething about Senfation in general;
not as defigning an entire and folemn Difcourfe of
that Subject, but becaufe the particular remark I
am about to make, is neceffary to the Solution of
our prefent Difficulty. I obferve then, that That,
which, at leaft in fuch cafes as we are fpeaking of,
produces in the mind thofe perceptions, which we
call Senfations of outward Objects, is the Local
Motion, caus'd by means of their Action upon the
Outward Organs in fome internal part of the Brain,
to which the Nerves belonging to thofe Organs correfpond; and the diverfity of Senfations may be referr'd to the differing modifications of thofe internal motions of the Brain, either according to their
greater or leffer Celerity, or other Circumftances, as
our Friend Mr. *Boyle* has fomewhere exemplify'd in
the variety of Sounds; whereof fome are grave,
fome fharp, fome harmonious and pleafant, fome
jarring and offenfive; and yet all this ftrange variety proceeds from the variations of thofe ftrokes or
impulfes, which the Air, put into motion by fono-
rous Bodies, gives to the ear.

· To this it will be confonant, that as the Air, or
rather the mind by the intervention of the Air, is
differingly affected by a very grave found, and a very
acute one; though the former proceed from the
want

want of that Celerity of motion in the undulating
Air, which is to be found in the latter; which flownefs or imminution of motion, does, as fuch, participate of, or approach to, the nature of Reft : fo
in the fenfory of Feeling, there may, upon the
Contact of a Cold Body, be produc'd a very differing perception from that which is caus'd by the
contact of a Hot Body; and this, though the thing
perceiv'd, and by us call'd Coldnefs, confifts but
in a leffer agitation of the parts of the cold Body,
than of thofe of the hot Body, in refpect of our hands
or other Organs of Feeling.

And this leads me, for the farther clearing of
this matter, to reprefent to you, that fince 'tis
manifeft, that Bodies in motion are wont to communicate of their motion to thofe more flow Bodies they happen to act upon, and to lofe of their
own motion by this communicating of it : Since
this, I fay, is fo, if, for Inftance, a man take a piece
of Ice in his hand, the agitation of the particles
of the Senfory will, in good part, be communicated
to the Corpufcles of the Ice, which, upon that account, will quickly begin to thaw; and the contiguous parts of the Hand lofing of the motion they
thus part with to the Ice, there needs nothing elfe
to leffen the agitation they had before. And there
needs no more than this flackning or Decrement of
Agitation, to occafion in the mind fuch a new and
differing perception, as men have tacitly agreed to
refer to Coldnefs.

Eleutherius. It feems by this Difcourfe, *Carneades,* that you think, that Senfation is properly and
ultimately made in, or by, the Mind, or difcerning
Faculty; which from the differing motions of the
inter-

internal parts of the Brain is excited and determin'd to differing perceptions ; to some of which Men have given the names of Heat, Cold, or other Qualities. So that, according to you, if a confiderable Change or Variation be made in the moſt ordinary, or in the former motion or modification of motion of the parts of a Senſory, and conſequently of the parts that anſwer them in the Brain, new Senſations will be produc'd, whatever the cauſe of this Alteration be, whether Privative or Poſitive.

Carneades. You do not miſ-apprehend my thoughts, *Eleutherius*, and what you ſay gives me a riſe to illuſtrate this matter yet a little farther by obſerving, that the Senſories may be ſo accuſtom'd to be affected after a certain manner by thoſe external Objects, whoſe Operation on them is very familiar, or perhaps almoſt conſtant, that the Privation, or the bare Imminution of the wonted operation leaves the parts of the Senſory, for want of it, in a different diſpoſition from what they formerly were in ; which change in the ſenſory, if it be not too ſmall, will be attended by a perception of it in the mind. To declare and confirm this by an example, we may conſider, that though Darkneſs be confeſſedly a Privation of Light, and the Degrees of it, gradual Imminutions of Light ; yet the Eye, that is, the Perceptive Faculty by the Intervention of the Eye may well enough be ſaid to perceive both Light and Darkneſs, that is, both a Poſitive thing, and the Privation of it. And 'tis obvious, that the motion of a ſhadow, which is a gradual Privation of Light, is plainly, and without difficulty, diſcoverable by the Eye ; of which
the

the reason may be easily deduc'd from what I have been lately saying. And to shew you that there is on these occasions such a change made in the Organs of Seeing, as is visible even to By-standers, I shall need but to appeal to the Experiment of making in the day time a Boy or Girl look towards an enlighten'd Window, and then towards an obscure part of the Room; for when the latter comes to be done, you will plainly perceive, that for want of such a degree of Light as was wont to come in at the Pupill, and straiten a little that perforation of the Uvea; that round Circular Hole, or, as you know they call it, Apple of the Eye, will grow very manifestly larger than it was before; and than it will appear again, if the Eye be expos'd to a less shaded Light.

This observation may be seconded, by what happens to a man, when coming out of the Sun-shine, where the Sun-beams much contract his Pupill to shut out an excessive Light that would be offensive to the Organ, he comes presently into a dark room, where he must continue some time before he can see others as well as he is seen by them, whose Pupills have had time to be so inlarged, as in that darker place to let in light enough to make Objects visible to their Eyes, which are not so to his, whose Pupills are yet contracted by the Light they were but just before exposed to. To this I might add divers other *Phænomena*, explicable upon the same grounds; but I shall rather chuse to relate to you an uncommon Accident, which happening to eyes somewhat unusually disposed, do's more remarkably discover, what alteration Darkness, or a privation of Light, may have upon those Organs. I know

a very Learned man, who is no lefs ftudious of Ma-
thematicks, and other real parts of Knowledge,
than skill'd in thofe which are taught of the Schools :
This *Virtuofo,* who feem'd to me to have fomething
peculiar in his eyes, confefs'd and complain'd to
me, that if he come, though but out of a moderate
light of the open air, into a room that is any thing
dark, he does not only feel fuch an alteration as
other men are wont to do on the like occafion ; but
is fo powerfully affected by it, that he thinks, he
fees flafhes of fire before his Eyes, and feels a trou-
blefome difcompofure in thofe parts, that fometimes
lafts an hour or two together, if he fo long continue
there.

Eleutherius. I know not, *Carneades,* whether
after this, you will think it any great confirmati-
tion of your Opinion, that *Ariftotle* has fomewhere
this faying, that, *Oculus cognofcit Lucem & Tene-
bras.*

Carneades. I thank you, *Eleutherius,* for fo pertinent an Allegation, though not for my own fake,
yet for theirs that will more eafily receive a Truth
upon the Teftimony of *Ariftotle,* than that of *Nature.* And now, I hope, that *Themiftius* will con-
fent, that difmiffing the Argument hitherto examin'd,
we proceed to the next.

SECT. III.

Them. SInce you will have it fo, I fhall comply at prefent, and the rather, becaufe
not only I forefee there will be occafion to fpeak

I hope, be prevailed with by the Argument I am about to propose, which is so manifestly grounded upon Sense, that without denying that we do feel what we feel, we cannot deny Cold to be a Positive Quality. For thus *Gassendus* most convincingly argues; *Cùm per hyemem immittimus manum in labentis fluminis aquam, quod frigus in ea sentitur non potest dici mera privatio, aliudque prorsus esse apparet sentiri aquam frigidam, & sentiri non calidam. Et fac eandem aquam gelari, sentietur haud dubiè frigidior, an dices hoc esse nihil aliud quàm minùs calidam sentiri? Atqui calida jam antea non erat, quomodo ergo potuit minùs calida effici?*

Carnead. I will not say, *Themistius*, his Argument is not specious, but you, perhaps, or at least *Eleutherius*, will not affirm it to be more than specious, if you please to consider with me two or three things that I have to suggest about it.

And first, to shew, *Themistius*, that, whatever he was just now intimating, Experimental Philosophers do not prefer the immediate Impressions made on the Senses to the dictates of Reason, though they think the Testimony of the Senses, however sometimes fallacious, much more informing than the Dictates of *Aristotle*, which are oftentimes (and that groundlesly) repugnant to them; I will represent to you, that the Organs of Sense, consider'd precisely as such, do onely receive Impressions from outward Objects, but not perceive what

what is the caufe and manner of thefe Impreffions, the Perception properly fo called of Caufes belonging to a fuperior Faculty, whofe property it is to judge whence the alterations made in the Senfories do proceed, as may eafily be proved, if I had time and need to do fo, by many Inftances, wherein the Senfes do, to fpeak in the ufual phrafe, mif-inform, and, as far as in them lies, delude us, and therefore muft be rectified by Reafon. As when the Eye reprefents a ftraight Stick, that has part of it under water, as if it were crooked; and two Fingers laid crofs over one another, reprefent us a fingle Bullet or a Button roll'd between them, as if there were a couple: So that 'tis very poffible (for I forbear faying 'tis true, having not yet proved it,) that though the Senfory be very manifeftly and vehemently affected upon the contact of cold Water, or other cold Bodies, yet the caufe of that impreffion or affection is, and may be judged and determin'd by Reafon to be, other, than that which the Senfe may to an inconfiderate perfon fuggeft. As when a Child, or one that never heard of the thing before, firft fees a Stick, whereof one part is in the Air, and the other under Water, he will prefently, but erroneoufly, conclude that *Phænomenon* to be caufed by the Stick's being crooked or broken.

Next we may confider, that Senfations may in divers cafes be made, as well from alterations that may happen in the internal parts of the Body, as from thofe that are manifeftly produced in the external Organ, by external Objects and Agents; as may appear by Hunger, Thirft, the Titillation of fome parts of the Body, barely upon Venereal

C   thoughts,

thoughts, and (which belongs directly to our present Argument) the great Coldness that we have
known Hysterical Women complain of in their
Heads and Backs, and the great and troublesome
degree of Cold, which we every day observe upon
the first invasion of the Fits of Agues, especially
Quartans ; which troublesome symptomes, that
sometimes last for several hours, are therefore commonly called the Cold Fits.

And now it would be seasonable for me to call
upon you to remember (and add to what I have
now said) that which at the beginning of our conference I took notice to you of about Sensation in
general ; if I did not presume that those things are
yet fresh enough in your memory, to allow me to
proceed directly to answer the Objection, which
I shall do, though not like a *School-man,* yet like a
Naturalist, by giving an account of the proposed
Phænomenon, without having recourse to that *Hypothesis* which 'tis urged to evince.

I observe then, that though in the respective
sence above-mention'd, Water, wherein the Objection supposes the hand to be plunged, be cold,
in regard its parts are less agitated, than the Spirits
and Bloud harbour'd in the Hand; yet in a Philosophical sence, it is not quite destitute of Heat,
since 'tis yet Water, not Ice, and would not be a
Liquor, but by reason of that various agitation of
its minute parts, wherein fluidity, a Quality essen-
tial to Liquors, consists.   Upon the score of this
respective Coldness of the Water, the Hand is
refrigerated ; for the Spirits and Juyces of that
Organ meeting in the Water with Particles much
less agitated than they are, communicate to them
some

fome part of their own Agitation, and thereby lofe it themfelves, upon which Decrement of wonted Agitation, fuch a change is made in the Senfory, and, (though not fo manifeftly) in fome other parts of the Body, as is perceived by the Animad-verfive Faculty under the Notion of Coldnefs ; Senfation, (whatever obfcure Definitions are wont to be given of it) being indeed an Internal Per-ception of the changes that happen in the Senfo-ries.

And if now, as the Objection fuppofes, the Water wherein the hand is plunged comes to be more refrigerated than before, the Spirits, Blood, and other parts of the hand, finding the Aqueous Corpufcles more flowly moved than formerly, muft, according to the Laws of Motion, (according to which a Body that meets another much more flowly moved than it felf, communicates to it more of its motion than if 'twere lefs flowly moved,) transfer to them a greater meafure of their own motion, and confequently themfelves come to be deprived of it : And upon this increafe of the flownefs of motion in the parts of the hand, there follows a new and proportionable perception of the Mind, and fo, a more vehement fenfation of Cold. But though it be not to be admired, that the bare flownefs of motion in the Object fhould be difcern-able by Senfe, albeit it feems to participate of Reft, which with you paffes for a Privation, fince the Ear perceives when a Voice grows faint, and when a fharp Sound degenerates into a flat one ; and we can perceive by the hand (abftracting from Heat and Cold) the celerity or flownefs of Bodies that in their paffage ftrike upon it , as for inftance, of

Winds,

Winds or streams; yet this is not the only thing I think fit to be taken notice of on this Occasion. For, I consider farther, that besides the most consistent and stable parts of the Hand, there are from the Heart and the Brain fresh blood and spirits continually transmitted to the Hand; and the former of these, the Blood, is, according to the Laws of its Circulation, and after it has received a great change in the much refrigerated Hand, carried back through other parts to the Heart; whence it is in the same Circulation distributed to the whole Body. To which may be added, that when the great refrigeration of the Hand happens, external Agents may contribute to the Effects of it, as I shall by and by have occasion to shew.

If then you please to remember, that upon the turning ones eye to the dark part of a room less inlighten'd than the Window, though Darkness be but a Privation, and though the Obscurity of that part be not absolute, but consist only in a less degree of Light; yet the action of the Spirits and other parts of the Body is so changed upon occasion of the Light's acting more faintly than was usual upon the Organ, that the Pupill is immediately and manifestly dilated, and in some cases, as in that which I mention'd to you of a Learned Man, much considerabler Effects ensue; you will not wonder, that, where not only the Spirits, but the Blood, (whence those Spirits are generated) that circulates through the whole Body, and upon whose Disposition all the other parts so much depend, is very much disaffected, there should be felt a great alteration in the Hand, which is the most immediately expos'd to the action of the cold Water.

And

And for the Reasons newly given, it ought to be as little strange, that in other parts of the Body, the disorder'd and not circulating Blood should have its wonted action on them considerably alter'd; since the more stable parts, and especially those external ones that are most expos'd to the Cold, have their pores straiten'd, and consequently their Texture somewhat alter'd; on the same occasion on which the wonted agitation of the Spirits with the Particles that compose the Blood, is notably lessen'd. And that such Causes may produce great Effects in a Humane Body, you will be more prone to admit, if you consider the disorders that happen in the cold fit of an Ague, and oftentimes upon the shutting up of those excrementitious steams that are wont to be discharged by insensible Transpiration; to whose being stop'd in the Body by the constriction of the Pores, which chiefly happens through Cold, some Learned Physicians, especially the famous *Sennertus*, impute the cause of most Feavers, as indeed Experience it self does but too frequently shew it to be guilty of many.

Philoponus. I confess, *Carneades,* you have said some things that I thought not on before; but yet *Gassendus*'s Argument seems to be such, that I fear 'twill be hard to hinder many from saying, That if Cold be but a Privation of Heat, 'tis a Privation of a strange nature: For, it may be introduc'd into Bodies that were not Hot before, nay, in some cases, into such as are naturally Cold, and also by consequence must have been put into a preternatural state to be at any time Hot.

Carneades. This Objection, *Philoponus,* being in
effect

effect so much the same with that of *Gassendus*, that it differs from it but in the dress you give it, 'twill scarce require a peculiar and distinct answer; and therefore, as soon as I have reminded you of the Distinction, that we have formerly made of the Vulgar and Philosophical sence of the word *COLD*, I shall need to alter but a little what I said before, by telling you, that since Fluidity consists in the various agitation of the insensible Corpuscles of a Liquor, and that Heat consists in a *tumultuary*, but a more vehement agitation of the insensible parts of a Body, and so, that Hot Water scarce differs otherwise than gradually, from that which is cold to Sense; if Cold be taken in the larger and Philosophical sence, it may well be said, that as long as Water retains the form of Water, and so continues to be a fluid Body, though it may be very cold to the Touch, yet it is not absolutely or perfectly cold, and therefore is capable of a farther degree of coldness, which it receives when brought to Congelation: for till then it was not destitute of those agile Corpuscles, that were requisite to keep it fluid; and till then, *Gassendus* himself must acknowledge, that it was not absolutely or perfectly cold; because He, as you may remember, did in his former (but lately mention'd) Argument ascribe the Glaciation of Water to the invasion of those that he calls Corpuscles of Cold.

Eleutherius. Give me leave to add, *Carneades*, that 'tis not every Glaciation it self that brings Liquors to be perfectly Cold in the Philosophical sence of that expression, and quite expells or subdues all the agile Particles that were in the Water before 'twas turn'd into Ice. For, I think, that to
effect

effect this change, 'tis sufficient, that so many of these restless Particles be destroyed or disabled, that there remains not enough of them to keep the Water in a state of Fluidity, so that the surplusage may yet continue in the frozen Liquor, and whilst they are there, perform several things, as the making it evaporable in the Air, and even odorous, and by *their* recess or destruction the Ice may grow yet more cold. And *as* this Notion suits very well with the differing degrees of hardness, that we find in differing portions of Ice, sometimes upon the account of the matter, (as frozen Water is harder than frozen Oyl,) and sometimes upon that of the different degrees of Cold in the same Water or other matter, (as our Friend somewhere observes;) *so* it may be highly confirmed by an Experiment I saw him make, but that is not yet published.

The summ of the Experiment was this; That he first put an Hermetically seal'd Thermoscope into a Glass broader at the top than at the bottom, and greas'd the inside with Tallow, that Ice might not strongly stick to it. In this Glass was put Water, more than enough to cover the ball of the Instrument; and that Water being warily frozen, notice was taken, whereabouts the tincted Spirit of Wine rested in the Stemm; after which, the Instrument and the Ice being removed into the open Air, upon an exceeding frosty morning, the Ice was taken off from the ball, and presently after, the tincted Liquor, as the maker of the tryal expected, subsided a pretty way (the length of the Instrument considered) below the former mark; which argued that he rightly guess'd, that such a degree of Cold as is sufficient to turn Water into Ice, may not produce

a Boa Body perfectly Cold; this Ice it self keeping the inclos'd ball, in a sence, warm, by fencing off the Air, which, at that time, (even in our temperate Clime) by the Effect appear'd to be colder than the very Ice. And, me thinks, it may strengthen *Carneades's* Discourse, to represent, that there is no sufficient cause, why many things that are reckon'd among Privations or Negations by the Peripateticks themselves, as well as Cold is by *Carneades,* may not admit of degrees; as may be exemplified by Deafness, Ignorance, and divers other things. And to bring a case, not very unlike that under consideration, we may take notice of a total Eclipse of the Moon, which you know alwayes happens when she is at the full. For Darkness in the Air being acknowledged to be a Privation or Negation of Light, when the Earth interposed between the Moon and the Sun has Eclipsed her, for instance, nine digits, (as Astronomers speak,) Men generally complain of darkness in the air, though there remain a considerable part of the *Discus* or the Hemisphere of the Moon obverted to us yet inlighten'd by the Sun; but when the interpos'd Earth proceeds to cover the remaining three digits, and so makes the Eclipse total, the darkness also is said and esteem'd to be much increas'd: Nor would men otherwise be perswaded, though *Themistius* should tell them, that the Air cannot have grown darker, though it were dark before, and indeed though the Air was more and more darken'd in proportion to the increase of the Eclipse, yet it was never compleatly darken'd 'till it became total. But I fear I dwell too long upon one Argument.

SECT.

SECT. IV.

Elen. LEt me therefore, *Carneades*, fumm up what I take to be your Doctrine, and tell thefe Gentlemen, that I think you do not look upon the Senfation of Cold as a thing effected by an intire Privation properly fo called and confider'd as fuch, but that according to you that flown-nefs of motion in the Particles of cold Water, which the Hand finds when 'tis thruft into that Liquor, does occafion the Spirits and the Corpufcles of the Blood to part with to thofe of the Water a confiderable fhare of their own furplufage of agitation, whereby they lofe it themfelves, upon which is confequent a Perception of this change made in the Hand, which, if it be very great, is alfo frequently accompanied with fome fenfible change in other parts of the Body, occafion'd chiefly by the frequent returns of the circulating and highly refrigerated Blood to the Heart, whence 'tis difperfed to the whole Body. According to which Doctrine, the Senfation of Cold is but a perception of the leffen'd Agitation of the parts of the Hand either ftable or fluid, efpecially of the Blood; which alterations are in great part produced, not by the coldnefs of the Water, as Cold is a Privation, but from the new modification of the action of the Blood and Spirits upon the Nervous and Membranous parts, the conftriction of whofe Pores concurrs to that Modification. And, if I do not mifunderftand your Opinion, *Carneades*, methinks it may be confirmed by this which I have known obferved served by experienc'd Chirurgeons, that by too strict Ligatures unskilfully made, an Arm, for instance, may be gangrenated; in which case, all the proper and immediate effect of the Ligature is but the *constriction* of the part, though that constriction being unusual and excessive, it proves the occasion of the mortifying of the Hand and Arm by hindring the free and usual access of the Blood and Spirits to that Limb; upon which, by the depraved action of the parts of the Body one upon another, and the concurrence of external Agents, there ensues a Mortification or Gangrene of the part, which, if due Remedies be not timely employed, is communicated to other parts and kills the Man.

Carneades. Whatever become of your Instance, *Eleutherius,* I thank you for your readiness to propose it in favour of my Hypothesis, which you will easily judge not to be much concern'd in the close of the excellent *Gassendus* his Arguments for the Positive Nature of Cold. For though these words of his ————

Themistius. You may save your self the trouble of naming them now, since, whatever they may seem to you, I profess I look upon them as containing a distinct Argument, which I shall therefore propose in its due place hereafter; but in the mean time, and before we leave the Argument you would have us dismiss, give me leave to remind you, *Carneades,* of some part of your former Discourse, and to take thence a rise to tell you, that you, who told us that we ought not to consider the Operations that Qualities have upon our own Sensories only, but also what they do to other Bodies,

will,

will, I hope, allow me to demand, how a Privation, or if you will, how an Imminution of Motion can produce the hundredth part of thofe Effects which we daily fee produc'd by Cold in the Bodies that are about us.

Carneades. I thought, *Themiftius*, I had intimated to you already, what might have prevented your Queftion; but fince I fee 'tis otherwife, you fhall not find me backward to explain my felf a little more fully. I do not pretend, that either an abfolute privation of motion in a Body, or a flownefs of motion in the parts of it, is, as fuch, the proper Efficient caufe of the Effects, vulgarly but unduely afcrib'd to Cold alone; for, in my opinion, Cold is rather the Occafion, than the true Efficient Caufe of fuch Effects, which, I think, are properly to be afcribed to thofe Phyfical Agents, whofe actions or operations happen to be otherwife modified than elfe they would have been upon the occafion of that imminution or flacknefs of Agitation which they meet with in cold Bodies, by occafion of which they are both deprived themfelves of the Agitation they communicate to fuch flow Bodies, and thereby act no longer as, were it not for that lofs, they would, and by a natural confequence of this change, which is made in themfelves, they do alfo, though lefs notably, modifie the action of other Bodies upon them : From which unufual alterations happening in a World fo fram'd as this of ours is, and govern'd by fuch Laws refpecting Motion and Reft as are obferved among Bodies, there muft in all probability refult many new, and fome of them confiderable, *Phænomena.* For though Quiefcent Bodies feem not to have any action which among corporeal

substances seems to be perform'd only by Local
motion; yet Bodies quiescent themselves may concurr to great Effects both by determining the motions of other Bodies this or that way, or by receiving their motion totally or in part, and so depri-
ving the formerly moving Bodies of it : Thus the
Arches of a Bridge, though immoveable themselves,
by guiding the water of the River that beats against
them, may occasion a rapid and boisterous stream,
capable to drive the greatest Mills, and perform
more considerable effects, though the River, before it met with them, ran calmly enough, as is
evident at *London* Bridge, especially when the Water is near a low Ebb. And now I have mention'd
Water, I will add, that though Water it self be not
a quiescent Body, but being a Liquor has its parts
in perpetual motion among themselves; yet since
that agitation exceeding slow in comparison of the
swiftness of a Cannon-bullet, in respect whereof
the calm surface of the Water participates of the
nature of a Quiescent Body, Bullets themselves
shot from out of Guns elevated but little above the
Level of the Water, (upon which score they make
but a very sharp angle with it;) these Bullets, I
say, do not unfrequently rebound from the Surface
of the Water, and consequently, even these so won-
derfully swift Bodies receive a new Determination
from it.

Eleutherius. One may add, *Carneades,* to your
Instances, that in a Tennis-Court the wall, against
which Balls are strongly impell'd by a Racket, contributes much to the mischief that those Balls do
often to By-standers in the Gallery, as the Wall,
though it self unmov'd, gives a new Determination

to the moving Ball, and by its refiftance makes it rebound or reflect at an Angle equal to that of the Balls incidence.   And this concurrence of the Wall to fuch Effects is the more evident, becaufe of this other circumftance, (which alfo befriends your Opinion,) that, if the impell'd Ball, inftead of hitting againft the Wall, hits againft the Net, this by yielding deprives the Ball of its *Impetus,* and hinders the reflection that would elfe enfue.

Carnead. You have, I confefs, fomewhat prevented me, *Eleutherius,* but yet not altogether : For though I was going to propofe the example of a Ball, yet 'twas in fomewhat a differing way; for I was about to propofe to *Themiftius* the example of a Ball, which if it be forcibly and perpendicularly thrown againft the hard Ground, has its Determination fo alter'd, that whereas it moved before towards the Centre of the Earth, it immediately, with almoft the like fwiftnefs of motion, tends directly upwards.   And if on the other fide you throw the Ball, not againft a hard, but againft a muddy piece of ground, it will not rebound, lofeing its own motion, by communicating it to the parts of the yielding Mudd; as may be in fome meafure illuftrated by the great commotion made in a fmall Pond of Water, when a Ball (or a round ftone) being but gently let fall upon the furface of it, has its motion thereby deaded, and transferr'd to the parts of the Liquor, which perhaps will be vifibly agitated at the remoteft brink of the Pond.

Eleutherius. Thefe Examples may conduce much to explicate your Doctrine, *Carneades,* but fince *Themiftius* himfelf was fo equitable a while agoe, as to allow you much time to defend such a Paradox as yours against *Gassendus*'s Argument, I shall with your leave (of which I doubt not) to the Examples already mention'd add this one more. Suppose upon a stream that runs through some Town (which is not very rare) there were built a number of differing Mills, some for the grinding of Corn, others for the Fulling of Cloth, others for the moving of Bellows to melt Oars and Metals; others for forging of Sword-blades; others for making of Paper, and others for other uses : And suppose that an Enemy coming to besiege this Town, should successfully imitate *Cyrus*'s Stratagem, when by suddenly diverting the course of *Euphrates* he took *Babylon*; would it not be consequent to this derivation of the Water into some lower place, and this ceasing of the Stream to run in its former Channel, that the action of all these Mills, by which so many differing operations were perform'd, must of necessity cease too? though the Besiegers do not produce this change by any positive and direct violence that they offer to the Mills, but onely by hindring them from receiving the wonted Impulses which were requisite to keep them in motion.

Carneades. I dislike not your Instance, *Eleutherius*, which yet will not altogether render useless what I was going to say about a Wind-mill, which will illustrate one part of my Doctrine, for which your Water-mill does not seem to have been intended. And that this Example may the better do so, I will suppose a Wind-mill to be built in some low place near the bank of your stream, which stream we will suppose to be lyable, as some others are, upon the falling of great and sudden rains upon

on

on the neighbouring hills, to overflow its banks, in
cafe the increafe of the Water be not then hindred
by the Wind-mills lifting up conftantly fome parts
of it, and conveying it away by Pipes or otherwife :
And then let us fuppofe, what really fometimes
happens, that the Wind fhould fo ceafe, that there
fhould not blow any wind ftrong enough to move
the fails for a great while together; will it not
hence manifeftly follow, that by reafon of this abfence of the Wind, which abfence has the nature of
a Privation or Negation of a Stream-like motion in
the Air, not only there will be a ceafing of thofe
Effects and Operations whatever they were, that
were wont to be perform'd within the Mill it felf,
but alfo there will be a durable intermiffion of that
main work of the Mill whereby it carried off fuch
a quantity of Water ; which work ceafing with the
Wind, whilft the flowing in of the Water does not
ceafe too, but continues as formerly, the ftill-in-
creafing Water muft bear down or overflow its
wonted Banks or other Boundaries, and by its un-
ruly effufions drown the neighbouring parts, and
produce the Diforders, that is, the new *Phænomena,*
naturally confequent to an Inundation made by fuch
a quantity of Water.   And if the Water conveyed
away by means of the Mill through Pipes or Chan-
nels were employed to water Grounds, or other
particular ufes, the growth or fertility at leaft of
the Vegetables that Water was requifite to nourifh, or the other ufes to which it was neceffary,
muft confequently be much, if not totally, hindred.

Philoponus. I know not whether we may not
refer to the Subject of your Difcourfe, what may
be

be obferv'd in *Paralytick* affections, where a little Vifcous or Narcotick Humour obftructing or otherwife difaffecting one part of a Nerve, though its proper and immediate action be only to hinder or weaken the Spirits, that were wont, in competent plenty, to pafs freely along the Nerve to the Mufcles whereto it leads ; yet the action of the other parts of the Body and the Relaxation of the Fibres do oftentimes produce a tremulous motion in the Limbs, and particularly the Hands ; and fometimes alfo the Mouth, Neck and other parts, are drawn awry in an odd and frightful manner.

Carneades. Though I approve of *Philoponus*'s fancy, yet I think a more quick and notable Inftance to the fame purpofe may be taken, from what happens to Birds, and Rats, and Cats, and fuch kind of warm Animals , in Mr. *Boyle*'s Engine. For *as* the Air by the agitation of its parts, or that of fome Ethereal fubftance that pervades it, entertains the fluidity of Water and other Aqueous Liquors; and when that agitation is hinder'd or too much leffen'd, Water ceafes to be fluid, and upon that divers Violent Effects enfue, wont to be afcrib'd to Glaciation : *fo* the bodies of warmer Animals, having been born in the Air , and perpetually expos'd to the action of it , (though that be feldome heeded) when being plac'd in the Receiver of the Air-pump, and by the operation of that Inftrument, which withdraws the former Air and keeps out the new, the Air that was wont continually to act upon them, is kept from doing fo any longer, though this abfence, or not touching of the Air, be but a privative or negative thing, yet by reafon of the ftructure of the *Animal*, his Spirits

rits and Humours, affifted by the concourfe of more general Caufes, are brought to act fo differingly from what they were wont to do, that the Blood and Juyces fwell, the Stomach vomits, the *Animal* grows faint and ftaggers, the Limbs, and at length the whole Body are convulfed, the Circulation is ftopp'd, and at laft the whole *Animal* kill'd ; and all this done in a very few minutes of an hour, without the vifible intervention of any pofitive A-gent.

Eleutherius. What you fay, *Carneades,* concern-ing the quick and violent Death of warm *Animals* in Mr. *Boyle's Engine,* puts me in mind of an Ex-periment I faw made in that Inftrument upon cold *Animals,* which, methinks, may well illuftrate the Comparifon we lately employed of a Wind-mill. For *as* thofe great artificial Engines lofe their Mo-tion, and the Operations depending on it, if that Stream of Air, we call the Wind, be held from keeping them going ; *fo* Infects and fome other cold *Animals* have their differing motions fo de-pendent upon the contact of the Air, that, as foon as ever they are deprived of it (by the Engine we are fpeaking of) divers forts of them will lye movelefs as if they were dead; and I have known feveral of them that were put in together, continue in that ftate for many hours, as long as it pleas'd our Friend to with-hold the Air ; but when once He thought fit to let a Stream of Air enter the Receiver, thefe feemingly dead Animals, as Worms, Bees, Flyes, &c. like fo many little Wind-mills of Nature's (or rather her great Author's) making, were fet a moving in various manners (as creeping, flying, &c.) fuitable to their differing *Species.*

D   *Carne-*

Carneades. So that to summ up in a few words the Result of these Instances, and the rest of the past Discourse on the same Subject, it appears by what has been said, that the Effects undeservedly ascrib'd to Cold, need not in our Hypothesis be referr'd to a Privation, but to those positive Agents or active Causes, which by their own nature are determin'd to act otherwise on, or suffer otherwise from, one another, in cases, where there is a great hindrance or ceasing of wonted agitation, than where there is not.

SECT. V.

Themist. IT may perhaps now be time to put *Carneades* in mind, that, in what he has been discoursing all this while, he has propos'd Answers but to a couple of *Gassendus's* Arguments, and left the rest untouch'd.

Carneades. I should readily grant, *Themistius,* that I have dwelt too long upon so few Arguments, if I did not hope, that by fully answering Them, and giving the Company a particular account of my Notions concerning Cold, I might very much shorten and facilitate the remaining part of my Task, which engages me to return Answers to the other Arguments you speak of, the grounds of solving which, I think, I have already laid in the past Discourse. And therefore you may go on to propose the next Argument of *Gassendus,* as soon as you please.

The

Themiſtius. And I ſhall do it, *Carneades*, in that Learned Man's own words, which I well remember to be theſe: *Fac manum immitti in aquam nunc calidam, nunc frigidam;* Gaſſend.*Lib.* 6. *Cap.* 6. *quamobrem manus intra iſtam, non intra illam refrigeratur? an quia calor manûs intra frigidam retrabitur, manúſque proinde relinquitur calida minús? At, quidnam calor refugit, quod intra frigidam reperiatur? nonne frigus? At ſi frigus eſt tantùm privatio, quidnam calor ab illa metuit? Privatio ſanè nihil eſt, atque adeò nihil agere, unde ejus motus incutiatur, poteſt.*

Carneades. This Objection, *Themiſtius*, may indeed puzzle many School-Philoſophers, but will eaſily admit an anſwer in my Hypotheſis. For that does not oblige, or ſo much as tempt, *me* to aſcribe (as a *Peripatetick* would do,) to a meer Quality, (for ſuch is Heat,) both a knowledge of its danger, and a care and skill to preſerve it ſelf from its Enemy, the Cold, by a retreat inwards. For, agreeably to what I lately delivered, 'tis obvious for me to explicate the *Phænomenon* thus : When a man puts his Hand into warm Water, the agitation of the Corpuſcles of that Liquor ſurpaſſing that of the Spirits, Blood, and other parts of his Hand, cannot but excite in him a ſenſe of Heat ; but when he puts the ſame Hand into cold Water, the caſe ought to be much altered, not by any imaginary retreat of the Spirits, but the communication of motion by other parts to the ſurrounding Water , by which means there muſt be in the Hand a great leſſening of the former agitation of its parts, the perception or ſenſe of which decrement of motion is that which we call the Feeling of Cold.

Eleu-

Eleutherius. I think indeed , *Carneades* , that
though this Argument may be confiderable againft
thofe that the Learned framer of it might have in
his Eye, it is but invalid againft you.   But can you
as well decline the force of that other Objeftion,
which *Gaffendus* more infifts on, and which feems
as direftly to oppofe *you* as any other Adverfaries
of his Hypothefis?

Themiftius. I prefume, *Eleutherius*, you mean
that cogent Argument, which *Gaffendus* propofes
and profecutes more fully than the reft, deducing
it from the way of artificially freezing Water by a
mixture of Snow and Salt, placed about the outfide
of the Glafs that contains the Liquor. For from this
practice he rationally concludes, that fince this
frigorifick mixture is through the Glafs able to freez
the Water into Ice, it may as juftly be affirm'd to
act by Corpufcles of Cold, as Fire can be to act
by Calorifick Corpufcles, when kindled Coals, placed on the outfide of the Glafs , make the contained Water boyl. And this cogent Argument will,
I hope, prove the more fatisfactory to *Carneades*,
fince 'tis not drawn from what he would call a di-
fputable Peripatetick Notion , but from the fame
Quiver, whence he affects to take his Shafts, *Experience* it felf.

Carneades. I freely acknowledge, Gentlemen,
this Argument to be very plaufible ; but that it is
clear and cogent, I muft not grant, till I be better
fatisfied that it is fo.

And, I fhall fcarce think it as evident, that Ice
and Salt act by a Pofitive Quality, as that burning
Coals do fo, though Cold feems as well to be produc'd by the former, as Heat by the latter. For

innumerable Experiments ſhew, that *Heat*, in the
Fire eſpecially, is a Poſitive Quality, conſiſting in
a tumultuary and vehement agitation of the minute
parts of the Body that is ſaid to be hot, and producing alſo in the Bodies that 'tis communicated to, a
local motion, which is manifeſtly a poſitive thing.
This is ſo evident, in the heating of Bodies by
mere attrition, the ſmoaking and melting of divers
Bodies in the Sun-beams (eſpecially at fit times of
the day and year,) the ſudden boiling and diſſipa-
tion of Water, Oil, *&c.* dropt on a red-hot iron, and
many other obvious inſtances, that 'twere a needleſs
work to go about to prove it, eſpecially ſince both
Themiſtius's Peripateticks, and *Gaſſendus* himſelf,
who ſo often diſagree about other things, agree
in confeſſing that Heat is a Poſitive Quality.

　Themiſtius. But remember, *Carneades*, that the
grounds on which they do ſo, are the ſame, on
which *Gaſſendus* juſtly builds the Propoſition, that
Cold alſo is a Poſitive Quality.

　Carneades. I did not forget that, *Themiſtius*; for
I was about to ſubjoyn to what I laſt ſaid, that 'tis
evident not onely by the confeſſion of my Adverſa-
ries, but by that (which to me is much more conſiderable) of Nature her ſelf, proclaiming it in the
Inſtances I juſt now mentioned, that Heat is a Poſitive Quality; whereas that Cold likewiſe is ſo,
does not appear to me by the Experiment of Artificial Congelations.　For, in this all that is clear in
matter of fact is, that Snow or beaten Ice and Salt
are put about a Veſſel full of Water or other Aqueous Liquor, and that, within a while after, this
Water begins to be turn'd into Ice ; but, that this
Glaciation is perform'd by ſwarms of atoms of Cold,

　　　　　　　　　　　　　　that

that permeating the Glass, invade and harden the Liquor, is not perceiv'd by Sense, but concluded by a Ratiocination, the cogency of which I am allow'd to examine without affronting the certainty of Sense; *that* not being concerned in the case. If then an intelligible way can be proposed of fairly explicating the *Phænomenon*, besides that insisted on by *Gassendus*, the objection drawn from this Experiment against my Hypothesis will be invalid. And such an Explication Monsieur *Des-Cartes* ingeniously gives in his Meteors : *Quia*

Lib. Meteor.
Cap. 3.

Materia Subtilis (sayes he) *partibus hujus aquæ circumfusa crassior aut minus subtilis, & consequenter plus virium habens, quàm illa quæ circa nivis partes hærebat, locum illius occupat, dum partes nivis liquescendo partibus Salis circumvolvuntur. Facilius enim per salsæ aquæ quàm per dulcis poros movetur, & perpetuo ex corpore uno in aliud transire nititur, ut ad ea loca perveniat in quibus motui suo minùs resistitur : quo ipso materia subtilior ex nive in aquam penetrat, ut egredienti succedat, & quum non satis valida sit, ad continuandam agitationem hujus aquæ, illam concrescere finit.*

Philoponus. I leave *Themistius* to consider, whether this Explication be without Exception; but I confess it is not without Analogy, and that even amongst the four first Qualities themselves. For when we *Chymists* have a mind to dry (for instance) the *Calces* or Precipitates or other Powders, from which we have filtrated the Liquors we employ to wash or dulcifie them, 'tis usual either to put the Filters, wherein these Powders remain almost in the form of Mudd, or to spread the stuff it self upon brown Paper or pieces of Brick or

Chalk,

Chalk, which much haften the exficcation of the things laid upon them, not by any drying Particles which they emitt into the foft fubftances, but by imbibing the fuperfluous parts of the Liquor, and thereby freeing from them the Subftances to be dry-ed. And, I remember, I have feen our Friend Mr. *Boyle*, by immerfing a piece of foft crumb of bread into an actually cold Liquor, that would ha-ftily imbibe its Aqueous Corpufcles, and dry it in a minute or two of an hour fo as to make it feel hard.

Eleutherius. Thefe inftances bring into my mind another Chymical Experiment, that I have feen made by the fame Gentleman, which was; That by putting into weak Spirit of Wine a fufficient quantity of Salt of Tartar, he quickly defleamed the Spirit without Diftillation, or fo much as Heat. And this will the better illuftrate the *Cartefian* Ex-plication, becaufe 'tis manifeft by the change that will be made of the moft part of the Salt of Tartar into a Liquor that will not mix with the now de-fleamed Spirit of Wine, that the reafon of the Operation is, that the Aqueous Particles of the Phlegmatick Spirit, finding, it feems, more conve-nience or facility to continue their motion among the Fixt Corpufcles of the Salt, than the Vinous ones of the Spirit, pafs into the Alkaly and diffolve it; and thereby defert the Liquor through which they were diffufed before. And I know another Saline body, that fo unites with Water, as not to be, by the Eye, diftinguifhable from it, and yet is of fuch a Texture, that Water is fo much lefs difpo-fed to mingle with it than with Spirit of Wine it felf, that it will forfake the Body it kept in agitation, to pass into this Spirit; and so leave that which it kept in the form of a Liquor before, to appear in the form of a consistent Body; which instance comes somewhat nearer than the former to the Experiment of Glaciation.

Carneades. Though what you have recited, Gentlemen, be not unwelcome to me, yet, I think, I can propose you an Experiment fitter to dilucidate the *Cartesian* Explication. For, I remember, that our common Friend, having a mind to shew, that a small proportion of agile matter, invisibly diffus'd through a Body that would be otherwise consistent, may bring it to, and keep it in, the state of Fluidity; devised and shewed me the following Experiment. He took Camphire broken into small bits, and casting a convenient quantity of it upon *Aqua fortis,* suffer'd it to float there, till without Heat the Camphire was dissolv'd into a Liquor, and it look'd and felt like an Oyl, which, though shaken with the *Aqua fortis,* would emerge to the top again. If this Oyl were kept well stopt, that the Spirits of the Menstruum might not evaporate, it would (as he affirm'd tryal had taught him) continue long fluid, he having sometimes kept it a year or two or more. And that 'tis the agile Spirits of the *Aqua fortis* that keep the Camphire fluid, he has made probable by divers things that I must not now stay to recite. And that the quantity of these agile Particles is but small, I am induc'd to think by this, among other things, that when I have made a small parcel of but moderate *Aqua fortis* turn a pretty proportion of Camphire into Oyl, & separated that Oyl from it, I could, by casting fresh Camphire on the same *Menstruum,* reduce that also into the form of Oyl.

Oyl. Now, that thefe Fluidifick Spirits (if I may
fo call them) are not fenfibly warm (no more than
the *Cartefian Materia Cœleftis*) in Water, is manifeft to the Touch : And whereas I at firft fufpected, that the reafon, why the pouring of this Oil into
Water doth prefently reduce it into Camphire again,
might be the coldnefs of the Water ; I after thought,
upon a farther information, that the reafon rather
was, that the Nitrous Spirits being difpos'd to pafs
out of the Oyl into the Water, this Liquor readily
imbib'd and diluted them, and confequently difabled
fo many of them, that thofe that remain'd could not
do their former work any longer : fince he had tryed purpofely, that the Reduction of the Oyl into
Camphire would prefently be made, though that
Liquor were not pour'd into Cold Water but Hot ;
fo that the agitation, that it received from the particles of the *Menftruum*, though not to our Touch
fenfibly warm, was much more efficacious, than that
which it received from the Heat of the Water.

Eleutherius. I know not, whether befides the
Inftances that have been now propofed, one may
not alledge fuch an Argument alfo in favour of the
Cartefian opinion about Cold, as would not be in-
fignificant, though it fhould be made appear, that
Cold may fometimes be produced by or upon the
Emiffion of Corpufcles, that in fome fence may be
call'd Frigorifick. For there may be Corpufcles
of fuch a Nature, as to fize, fhape, and other at-
tributes, as to be fit to enter the Pores, and pierce
even into the inward parts of Water, and fome
other Bodies, fo as to expell the calorifick Corpufcles they chance to meet with, or to clog or hinder their activity, or on fome other account confi-
de-

derably to lessen that agitation of the minute parts, by which the Fluidity of Liquors and the Warmth of other Bodies is maintain'd. But, even in such cases, though the Agent and the Actions that produce Coldness, be Positive things; yet the Nature of Coldness it self may consist in a Privation. As when a man is kill'd by a bullet, his Death is effected by a Positive and even impetuous Action, and yet Death it self is but a Privation of Life. If also in a dark Room a man cast cold Water upon a burning Coal, though the Water act by its Positive Quality of moisture, and, by virtue of that, extinguish the Fire, and by that means destroy the Light, yet the Darkness that is consequent upon this action, is not a Positive thing but a Privation.

SECT. VI.

Philop. THe pause you here made, Gentlemen, makes me think it seasonable to put the Company in mind, that it begins to grow late, and therefore to call upon *Themistius* to produce what he has yet to alledge out of *Gassendus.*

Themistius. The Philosopher you have nam'd, has indeed another Weapon to destroy the Errour about Cold, which he confutes. And this Argument like a two-edged Sword that cuts on both sides, does not only confirm what he maintains, but destroy the chief objection that can be made by his Adversaries. The Argument I speak of he proposes in these terms : *Tametsi multa videantur ex sola caloris*

caloris absentia frigescere, nihilominus nisi frigus extrinsecus introducatur, non tam profectò frigescere quàm decalescere sunt sensenda. Esto enim lapis, lignum, aut aliquid aliud, quod nec calidum nec frigidum sit, id ubi fuerit admotum igni calefiet sanè; at cùm deinceps calor excedet, neque frigidum ullum circumstabit, non erit cur dicas ipsum frigefieri potius quàm minùs calidum fieri, redirève in suum statum.

Carneades. Whether this contain not a difpute *de modo loquendi,* I fhall leave the Company to judge by what I fhall return in anfwer to it. I fay then, that it feems to me, that there is in the Difcourfe an Obfcurity, if not an Ambiguity, though I am confident not affected by the Candid *Gaffendus.* But to anfwer as directly as I can; If we fpeak only of a Coldnefs as to Senfe, I fee not, why Water or Wood or any fuch Body that is heated by the Fire, may not upon its removal thence be faid to grow Cold and not barely to *decalefcere* in our Philofophers fence of that word. For the Heat and Coldnefs of Water, in reference to Senfe, confifting, (as I lately fhew'd) in this, that the Particles of it are more or lefs agitated than the Hand that is immerfed in it, they need nothing elfe to make the Liquor grow Cold, than fuch an imminution of the brisk motion of its Corpufcles, that they ceafe to be as much agitated as thofe of our Organs of Feeling: And if this already impair'd agitation be ftill more or more leffen'd, the Liquor will ftill grow colder and colder without the help of any Pofitive Caufe, 'till at length the agile parts, that kept it fluid, being quite expell'd or difabled, the form of the Liquor comes to be exchang'd for that of Ice.

Philoponus. But what say you to that part of *Gassendus's* Argument, where he proposes an Adiaphorous Body, which, when affected with an adventitious Heat, would not grow cold by the bare removal or cessation of that Heat, unless it were refrigerated by an Agent, that were positively and actively Cold?

Eleutherius. I say, *Philoponus,* this Supposition should not be made, and that I know of no such Adiaphorous Body. For since, as I have been oblig'd to inculcate, those Bodies must be Cold as to sense, whose parts are less agitated than those of our Hands, and consequently Metals, Stone, Wood, and other Solid Bodies, and also Water, Wine, and all other unmingled Liquors we know, being heated by the Fire, will grow cold again of themselves, because the adventitious motion ceasing by degrees, either upon the recess of the Igneous Corpuscles, or the imparting of the extraneous agitation to the Air or other contiguous Bodies, the Stone or Water, *&c.* will again have so much fainter an agitation, than that of a mans Sensory, as to be by him judged Cold: And because almost all the *Species* of permanent Bodies here below that are known, have in what is call'd their Natural state a less degree of Agitation of their Insensible Parts, than mens Organs of Feeling are wont to have, those Bodies may be said to be Naturally Cold, and therefore ought not to be suppos'd to be indifferent to Cold or Heat.

Philoponus. But whether or no Nature do really afford us an Adiaphorous Body; yet surely the mind is able to conceive one, and therefore *Gassendus* may be allowed to suppose such Bodies, and

Car-

Carneades may be oblig'd to anfwer what he argues upon that Suppofition.

Carneades. 'Tis one thing to propofe an Adiaphorous Body, as barely an intelligible or a poffible thing ; and another, to give Inftances of it, as *Gaffendus* has done in particular Bodies, in which that indifference is not to be found. And 'tis this laft kind of Suppofition that I difallowed in *Gaffendus*'s Argument. But if a Body fhould be propofed as Adiaphorous in reference to Heat or Cold, I might fay without prejudice to my Caufe, that if fuch a Body fhould be carried into a hot place, it might there grow warm ; and if it fhould be removed back again, and kept till it loft that new adventitious Heat, it might rather *decalefcere* than grow cold as to Senfe. But the reafon is, becaufe 'tis not every degree of imminution of Heat that is able to denominate a body Cold, but fuch a degree as reduces the parts of it to a fainter motion than is at that time in thofe of our Organs of Feeling ; and till this be done, or at leaft very near done, the propofed Body is ftill (if I may fo fpeak) in the ftate of Heat as to Senfe : Which laft words I add, becaufe that in reference to other Bodies it may then be notably refrigerated. As Lead that has but heat enough to keep it in Fufion, may, by the pouring on of fuch Water as to a mans hand would feel Hot, be brought to grow hard, which lofs of Fluidity is alfo the Natural Effect of Cold, though perhaps both the Metal and the Liquor be yet as to Senfe confiderably Hot.

Eleutherius. So that, according to you, none of the kinds of Bodies that are actually known in Nature, are Adiaphorous as to Senfe. On which occafion

fion let me note by the by, that the frequent Variations of Sense muſt render it but an uncertain ſtandard of Heat and Cold: And upon ſuppoſition, that there were an Adiaphorous Body in reference to our ſenſe ; yet it would not be ſo in reference to all other Bodies, or, in the phraſe of our *Verulam* ſpeaking of Heat, *in ordine ad Univerſum.* And for what remains, the controverſie grounded on *Gaſſendus's* Argument ſeems to be rather Verbal than Real, and may be determined or compoſed by ſetling the diſtinct acceptions of the words *Cold* and *Heat.*

SECT. VII.

Philop. WHerefore I wiſh , that we may not waſte the little time that is left us upon Niceties of no greater concernment ; and I think this ſhort time would be better imployed, if *Carneades* would be pleaſed to tell us a little more particularly, what he ſuppoſes to be the thing that with-held Mr. *Boyle* from delivering an opinion about the *Nature* of *Cold.*

Eleutherius. Yet me thinks 'tis but fair , that *Carneades,* who has all this while been confin'd to the anſwering anothers Arguments, ſhould now take his turn to propoſe his own.

Carneades. I find in each of your motions, Gentlemen, ſomething ſo equitable and ſo expedient, that I ſhall in part comply with both. And that I

may

may haften to do what *Philoponus* defires, I fhall do
no more than briefly point at two things that may
be alledged in favour of the Hypothefis I defend.
For if you reflect upon what we have already dif-
courfed, we may take notice of things there,
that will fcarce be well accounted for by being
afcribed to Pofitive Cold, but may be far better
explained agreeably to our *Hypothefis*. And I
muft add in the next place, that I, who fu-
ftained the perfon of a Refpondent, may pretend
to have fufficiently difcharged my Office, if I
have fhewn the invalidity of all the Opponents
Arguments; and 'tis his part who afferts a pofitive thing in Nature, to make it good, whereas he
that denies it, needs not alledge any other reafon
why he does fo, than the Authority of that juftly
received Axiom in Philofophizing, *Entia non funt
multiplicanda abfque Neceffitate*. And, I hope,
there will need no other Engine to demolifh an
ill-formed and prooflefs Opinion about Cold,
than an *Axiom* fo Solid and Efficacious, that
in the Opinion of almoft all the Modern Natu-
ralifts it has been able to abolifh fuch potent
and immenfe Bodies as the *Primum Mobile* it felf,
and a fuperior Orb or two, the leaft of which contained that Firmament, in comparifon whereof the
whole Earth is but a point. And not only fo, but
the fame Axiom has banifhed the Angels and
Intelligences from the Celeftial Orbs, that *Ariftotle* and his followers had affigned them to turn
about; or rather hath releafed thofe Noble and
Happy Spirits from the drudgery to which the
Philofophers of fo many ages had needlefly doom'd
them.

Elem-

Eleutherius. I the less distrust the validity of the Axiom you alledge, because I observe it to be the ground, on which is built a great part of the Reformation of Philosophy, that is introduc'd by the Moderns. For one of the main things that first moved considering men to seek for more satisfactory Opinions than those of the *Peripatetick* Schools, was, that these obtruded a great many Tenents in Philosophy, that were not only un-proved, but unnecessary to the Explication of the *Phænomena* of Nature; as 'twere not difficult to shew.

But I see *Philoponus* preparing to renew the motion he lately made, in which the shortness of time makes me now think it seasonable to joyn with him, I being no less desirous than he to know, what may be the motives of your Friend's declining to declare himself fully about the Nature and Cause of Cold.

Carneades. I have already intimated to you at the beginning of our Conference, that he is himself the fittest person to be addrest to for satisfying this inquiry. But not to be altogether silent on this occasion, I shall tell you, that, as far as I can guess, he waits till farther Tryals and Speculations have resolved him in some points, wherein he is not yet satisfied: For, being of a temper backward enough to acquiesce without sufficient Evidence, when the inquiry is difficult and the subject important; he seems to me to be kept in suspence, both by some Speculative doubts, and the *Phænomena* of divers Experiments, some of which are not deliver'd in his Book. It would be now improper to mention the

scruples

scruples and hesitancies they have occasioned in him ; though of those, I have heard him speak of, I shall name some Instances that occurr the most readily.   As I remember I heard him make inquiry, as to those that would have Cold produced by Corpuscles of Cold ; *Whether*, and on what account, those little fragments of Matter are cold ? *Whether* those frigorifick Particles, that must in multitudes crowd into Water to turn it into Ice, have Gravity or Levity, or are indifferent to both ? And how any of the three Answers that may be made to this Inquiry, will agree to some *Phænomena* that may be produced ? *What* Structure the Corpuscles of Cold can be of, that should make them frigorifick to that innumerable variety of Bodies they are said to pervade ? And *whether* the frigorifick faculty of these Corpuscles be loosable or not ? As also *whether or no* they be Primitive Bodies, and if it be said, they are not, whether there was not Cold in the World before they were produced, and whence that Cold could proceed ? And if it were said they *are* Primitive Bodies, he demanded, *how* it came to pass, that, by putting a certain factitious Body actually warm into Water that was also warm, (both which appeared by a good sealed Weatherglass,) there should presently be produced an actual Coldness (discernable by the same Thermoscope ?)   These, and I know not what other Scruples and Difficulties, suggested to him by his thoughts or his Experiments, were the things that I suppose prevail'd with a man of his temper to forbear for a while the declaring of his Sentiments about Cold, left the Event of some

E   farther

farther tryal should shew him cause to retract them.

Philoponus. What you have freshly intimated, *Carneades,* of Mr. *Boyle's* having other hesitations than those you have named and suggested by Experiments not published in his History, does, I confess, the more excite my Curiosity to have at least a taste of those perplexing *Phænomena.*

Carneades. You may easily guess, *Philoponus,* by what I have told you already, that you are not to expect a full satisfaction from me on this occasion. But yet, that your curiosity may not be frustrated, I shall venture to acquaint you with two *Phænomena,* which were, I suppose, none of the least motives of his backwardness to declare himself. But though some body perhaps thinks, that the grounds of solving these *Phænomena* and most of the newly recited Scruples, may be pickt out of some things that may already have pass'd among us in this Conference; yet because we have not now time to enter upon a discussion of this matter, I am willing you should suspend the debate, till we have occasion to meet another time; and therefore I shall now only acquaint you with a couple of Experiments, that he set down for a *Virtuso,* who was to solve the two main Problems suggested by them. The *first* whereof was, Whence Water should upon Congelation acquire so vast a force as he found it had to lift up great Weights and burst containing Bodies; though it seem'd by several Circumstances, that the motion of the Water is very much diminished when 'tis changed into Ice. And the *second* Problem is thus conceived; If, as a brisk agitation of a Bodies insensible parts produces

duces Heat, fo the Privation of that Motion is, as *Cardan* and the *Cartefians* would have it, the caufe of Cold ; whence is it, that, if certain Bodies be put together, there will be a manifeft and furious agitation of the fmall parts, and yet upon this conflict the mixture will not grow hot, but fenfibly and even confiderably Cold? The Narratives themfelves of the Experiments are too long to be now read over to you. And therefore I fhall leave the Paper, that contains them, among you, to be perufed at your leifure, between this and our next meeting, till when I muft bid you farewell : Only defiring you in the mean while to remember, that, as I have but acted a part impofed upon me in our paft Conference, fo notwithftanding any thing that I have faid in my affum'd Capacity, I referve to my felf the right of appearing as little preingaged as any of you at our next meeting.

F I N·I S.

... of that Mineral is, as
... Caufe, w would have it, the cruft
... it certain Bodies be
... and furious
... and yet upon this com-
... fenfibly and
... The Narratives them-
... are too long to be now
... I fhall leave the
... ing you, to be perufed
... and our next meeting,
I muft bid ... farewell: Only defiring
... man who is to remember, that, as I
... a part injoyded upon me in our paft
... fo ... underftanding any thing that
... is my utmoft Capacity, I referve to my
... as little prejudiced as

TWO PROBLEMS

ABOUT

C O L D,

Grounded on

NEW EXPERIMENTS,

And Propoſed

In a LETTER to a FRIEND,

By the Honourable *ROBERT BOYLE.*

TWO PROBLEMS

ABOUT

GOLD.

Grounded on

NEW EXPERIMENTS,

And sent in

A LETTER to a FRIEND.

By the Honourable ROBERT BOYLE.

To my very Learned Friend Mr. J. B.

SIR,

I Prefume that you will not be furpriz'd to be told, that I fend you the inclos'd Papers, not only that I might gratifie your Curiofity, but that you may by them be inabled to help me to fatisfie my own; and therefore I fhall accompany the Hiftorical Tranfcripts I made of the following Experiments, as I found them regiftred for my own Remembrance, with fome of the doubts fuggefted to me by fome of the *Phænomena* that occurr'd. But yet I fhall not trouble you with all the difficulties that at firft troubled me, but reduce the Exercife, I defire to give your fagacity, to the folution of two Problems. And I will begin with propounding that firft, which is grounded upon the laft of the two following Papers, becaufe though the Hiftorical part of that be much the longeft, yet the grounds of my *Quære* concerning it, will be much more briefly propos'd, the Experiment it felf naturally fuggefting this Problem; *How upon the mixture of two or three Bodies, fuch as thofe mentioned in the Paper, there fhould manifeftly*   **Problem. II.**

en-

enſue a great and tumultuary agitation of ſmall parts, and yet even during this conflict, not any ſenſible Heat, but *a conſiderable degree of Cold be pro-duc'd,* and that even in the internal parts of the mixture ?

The Inducements to make this Problem need not be far fetch'd, it being obvious enough, that, according to the Corpuſcularian Philoſophy, which you and I agree in, a brisk and various agitation of the minute parts of a Body is that, which makes it Hot both in reference to our Senſories, and to its operations on other Bodies. But I doubt, the riſe of the Problem is much more eaſie to be under-ſtood, than the Cauſe of the *Phænomenon,* about which I will not ask you, Whether one may not aſ-ſert, that Local motion is in its own nature a Gene-rical thing, which may be ſo diverſified by Cir-cumſtances, that one kind of Modification of it, as 'tis made in Corpuſcles of ſeveral ſizes and ſhapes, may be the cauſe of Heat, and another that of Cold ? Or elſe, Whether we may ſuppoſe, that Cold is a poſitive thing, and operates by real Cor-paſcles of Cold, which happening to abound, and yet to be lock'd up in the Bodies whoſe mixture I imploy'd, they are, by the great conflict that diſſolves the Texture of the Claſhing Salts, ſepa-rately put into motion and that in ſuch numbers, that though really there would be a Heat produc'd by the brisk and confus'd agitation of ſome of the parts, yet that Heat is not only conceal'd and check'd, but maſter'd by the over-powering opera-tion of the Frigorifick Corpuſcles. But to ask you about this or any other particular way of ſolving our *Phænomenon,* were to forget, that my aim is to

learn

learn not your opinion of this or that particular Conjecture or fancy about our Problem, but in general, how it may be beft refolv'd, and what you think to be the true Caufe of fo odd an Effect.

Having thus difpatch'd the little I had to fay about the Paper that fuggefted the fecond Problem, I will now fuppofe that you have read the *Phænomena* that contain the rife of the firft, to which I fhall proceed without farther Preamble, fince the Queftion or Problem, that thefe naturally call for, is, *Whence this vaft force of freezing Water proceeds ?*   Problem I.

For, the breaking of refifting Bodies being to be made by a violent Local motion, and Cold, according to the Judgment even of the Moderns, either confifting in, or at leaft being accompanied with, a Privation, or a great Imminution of Motion, it feems very difficult to conceive how Cold fhould make Water to exert fo wonderful a force. I know the Learned *Gaffendus* and divers other Philofophers teach us, that Glaciation is perform'd by the entring of fwarms of Corpufcles of Cold, as they call them, into the Liquor. But I much doubt, Whether from this Hypothefis a good Solution of our *Phænomenon* will be deriv'd, fince thefe Atoms of Cold feem not barely as fuch to make that Expanfion of the Water, which is requit'd in the Experiment by me recited. For I fee that though Water will b more and more refrigerated, according as the Air grows colder and colder, yet till it be brought to an actual Glaciation, all the fwarms of the Frigorifick Atoms in it, are fo far from expanding it, that they more and more condenfe it. And ever that degree of Cold which deftroys Fluidity, though it expands Water,

Water, does not do it merely by the multitudes of the Frigorifick Corpuscles that invade the Pores of the lately fluid Body, since pure Spirit of Wine and almost all Chymical Oyles, though expos'd to the same degree of Cold that turns Water into Ice, or as I have tryed, unto a far greater than is necessary to do so, will be but the more condens'd by those swarms of Particles. But, which is more considerable, I have carefully observ'd, that, besides common or exprest Oyls, Chymical Oyl of Aniseeds it self, being frozen or concreted by an intense degree of Cold, will not be expanded but notably condens'd, and accordingly grow specifically heavier than before. And this was one thing that kept me from expecting the removal of our Difficulty from the Ingenious Explication given of Freezing by the *Cartesians*, when they teach, that the Eel-like particles whereof they suppose Water to consist, are very remissly agitated, and their want of pliantness makes their Contexture less close; which yet seems not to agree with the lately mention'd Tryals. And though these Eel-like particles should lose all their flexibleness, though in that case it may probably be said, that they would take up less room than before, if nothing oppose their Expansion, yet it does not thence appear, how they should acquire so vast a power to expand themselves in spite of Opposition, as we have shewn Water by Freezing does acquire.

I did not hope to resolve our Problem by the help of a Vulgar Supposition, that well-stopp'd Vessels are broken in frosty weather *ob fugam Vacui*, since I found that Supposition to be erroneous by divers Experiments, some of which are mention'd in the History of Cold.   It

It feem'd lefs improbable, that fome affiftance to the folving of our difficulty might be given by two other things. Whereof the firft is, That, for ought I have yet obferv'd, no Liquor but Water, or that which participates of Water by having Aqueous Particles feparable from it, will be made to fwell by Cold; nor will Water it felf do fo upon every degree of Cold, but only upon fo great an one as actually turns it into Ice. And the fecond is, That upon the Glaciation of Water and Aqueous Liquors, we may obferve in the Ice many bubbles greater or fmaller intercepted between the Solid parts, and fuppos'd to be full of Air, (I fay fuppos'd, becaufe upon tryal I found them to have yielded but a fmall proportion of common Air;) which fuppofition, if true, would perhaps invite one to fufpect, that the Air contain'd in thefe bubbles might have an intereft in our *Phænomenon*; fince I have found by tryals purpofely made, that Air congregated into Vifible though not great portions, may exercife a confiderable Elafticity, which appear'd not whilft 'twas invifibly difperfed through the Water.

And if I did not fuppofe, both *that* you had taken notice, that there are wont to be numerous particles of fpringy Air difpers'd through the Pores of Water; and *that* you had confider'd, *whether* the want of pliantnefs occafion'd by Cold in the Aqueous Corpufcles, whilft they are yet agitated and brandifh'd by fome permeating matter; and *whether* upon the change of the Pores, that we may conceive to be made in freezing Water, either by the recefs of one fort of fubtil Corpufcles or the admiffion of another, or the clofer conftipation of the groffer parts, there may not be produc'd in Corpufcles, that compofe
Wa-

Water, (to fay nothing of the intermix'd Air, or the Concretions or the Coalitions occafion'd by the Cold,) a fpringinefs capable to make many little Bodies, endow'd with it, exert a great force againft the fides of the Veffel, that oppofe their joynt endeavour to expand themfelves: If, I fay, I did not believe, that thefe and the like fufpicions had occurr'd to you as well as to me, together with the difficulties wherewith each of them feems to be incumber'd, I would acquaint you with what thoughts and tryals occurr'd to me about thefe and the like conceits. But I not daring to think this could prove other than a needlefs work, I muft remember, that my bufinefs in this Paper is to propofe Difficulties, not the wayes of folving them ; it being from your Kindnefs and Sagacity, that thefe are as well expected as defir'd by,

S I R,

Your, &c.

AN ATTEMPT

To Manifest and Measure the

GREAT EXPANSIVE FORCE

OF

Freezing Water.

By the Honourable *ROBERT BOYLE*.

Freezing Water.

By the Honourable ROBERT BOYLE.

AN ATTEMPT

To Manifeſt and Meaſure the

GREAT EXPANSIVE FORCE

OF

Freezing Water.

COnſidering when I writ the Hiſtory of Cold, that though divers *Phænomena* might induce an attentive Obſerver to think, that Freezing Water had an Expanſive Force, yet I had not met with any that endeavour'd, or even propos'd, to meaſure it, whether becauſe they reflected not on it at all, or judg'd not the Force conſiderable; I who look'd with other eyes upon it, thought fit to repair that omiſſion, but was then ſo ill furniſhed with requiſites for doing it fully, that I remember I complain'd of it in my Hiſtory of Cold. And though, even afterwards, when the time of the year was favourable, I could not procure ſuch Accommodations as my deſign exacted; yet thinking an imperfect way of Meaſuring to be better than none, I preferr'd to the making no attempt at all the endeavouring to do what the least defective Inftruments, I could procure, would permit me, towards the making an eftimate by known Meaſures, of the Expanfive Power of Freezing Water. For though I did not expect, I ſhould be able accurately to define it; yet I hop'd I ſhould make ſuch an eftimate as to know that Force not to be, as one would think it, Faint and Contemptible, but very Great and Confiderable.

I remember on this occaſion, that to manifeft the Force of Freezing Water, I cauſed the Barrel of a ſhort Gun to have a ſkrew fitted to the Noſe of it, by which we might exactly ſtop it, as we did the Touch-hole another way; then filling the Barrel with common Water, and cloſing it accurately by the help of the ſkrew, we laid it in a conveniently ſhap'd Veſſel, wherein we incompaſs'd it with a Frigorifick Mixture (of Snow or Ice and Salt,) and in a ſhort-time we found, as we expected, the Barrel to be burſt, part of the Ice appearing along the gaping ſlit that had-been made in the Body of the Iron by the Freezing-Water, which by this Effect ſeem'd to emulate the juftly admir'd force of kindled Gun-powder. But the Deſign of this ſhort Paper tending not ſo much to *prove,* as (in ſome ſort) to *meaſure* the Expanfive Force of Water, I ſhall ſubjoyn the Tranfcripts of two or three Experiments, made chiefly for that purpoſe.

EXPERIMENT I.

[There was taken a ſtrong Cylinder of Braſs, whoſe Cavity was two inches in *Diameter*, into this was put a Bladder of a convenient ſize, with a
quanti-

quantity of Water in it, that the neck of the Bladder (which I had taken care to have oyl'd) being
ftrongly tyed, the Water might not get out into the
cavity of the Cylinder, nor be capable of expanding
it felf fome other way than upwards. Then into this
Cylinder was fitted a Plugg of Wood, turn'd on
purpofe, which was fomewhat lefs in Diameter than
the Cylindrical cavity, that it might rife and fall eafily in it. Upon the upper part of this Plugg was
laid a conveniently fhap'd flat Body, upon which
were plac'd divers weights to deprefs the Plugg,
and hinder its being lifted up by the Expanfion wont
to be made in Water that is made to freeze ; then a
Frigorifick Mixture being afterwards apply'd to the
Cylinder, it appear'd within half an hour or fomewhat more, by a Circle that had been purpofely
trac'd on the fide of the Plugg, where 'twas almoft
contiguous to the Orifice of the Cylinder, that the
Water in the Bladder began to expand it felf, and
about two hours after, having occafion to fhew the
Experiment to fome inquifitive perfons, the circle
appear'd to have been heav'd up in my eftimate a-
bout $\frac{3}{4}$, if not half, of an inch, notwithftanding all
the weights that endeavour'd to hinder the afcenfi-
on, though thefe weights amounted to 115 pound,
which were all the determinate weights we could
then procure, befides a brick and fome other things
that were eftimated at five pound more ; nor did I
doubt that a far greater Load would not have hindred its Expanfion.]

EXPERIMENT II.

[We took a Brafs Cylinder, whofe Dimenfions were three inches $\frac{8}{10}$ in Diameter, and in depth four inches. Into this we put a fine bladder of a convenient fize, almoft fill'd with Water, and ftrongly tyed about the neck ; upon this bladder we put the wooden plugg to ftop up the Orifice as much as was convenient, and upon the plugg we put a piece of a flat board for the weights to ftand upon. Thefe things being prepar'd, we convey'd the Cylinder with all that belong'd to it, fave the board, into a large wooden Bowl, where we applied to the Cylinder a good quantity of the Frigorifick Mixture, made with beaten Ice and Bay Salt; and having firft mark'd with a circular line the Edge or Contact, where the orifice or lip of the Cylinder touch'd the Plugg, we layed on the weights upon the board, and when by their weight they had deprefs'd the Plugg till the cover of it lean'd upon the Cylinder, we difpos'd our felves to attend the iffue of the Tryal. The event whereof was this, that when the action of the Frigorifick Mixture had produc'd fome Ice in the Water included in the Bladder, that Liquor appear'd to have dilated it felf ftrongly enough to begin to raife the Plogg with the fuperincumbent weights, and by degrees they were by the growing Ice rais'd till the mark, diligently made on the Plugg where the edge of the Cylinder touch'd it, was about a tenth part of an inch above the ftation it had before the Plugg had been deprefs'd. Then we took out the Bladder, and found the Cylinder of Water within the Bladder not to be

be wholly turn'd into Ice, but to contain fome
quantity of unfrozen Water in the parts about the
Centre, which Liquor, if we had not fo foon de-
fifted from the Experiment, (as for certain Reafons we did) might probably have rais'd the weights
fomewhat higher.   But as it was, the Ice in length
was but three inches and about ⅛, and yet fo fmall a
quantity of Ice fuffic'd to raife, befides the board
they lean'd on, as many weights of Lead as amounted to an hundred pound *Averdupois.*]

EXPERIMENT III.

[The day after the above mention'd Experiment
was made, to try yet farther the Expanfive force of
Freezing Water, the fame was reiterated after the
manner.above deliver'd, but with this difference,
that, having procur'd more weight, when the Plugg
was lifted up ⅛ or fomewhat better (which Plugg
began fenfibly to rife within half or three quarters
of an hour after the Frigorifick Mixture was applyed,) it was loaded with a weight of two hundred
pounds, and a fifteen pound piece of Lead, and
other Bodies, as Boards, &c. to lay the weights
upon, which being alfo weigh'd by themfelves came
to fifteen pound more, fo that the whole amounted
to 230 pound; and if the hundred pounds were
both of them, as their bulk and fhape invited us to
guefs, of that fort of weights which are call'd the
greater Hundred, containing an hundred and twelve
pound a piece, twenty four pound muft be added
to the famm, which would thereby be made up
254 pound.]

A NEW

EXPERIMENT

ABOUT THE

PRODUCTION of COLD.

BY THE

CONFLICT of BODIES,

Appearing to make an EBULLITION.

By the Honourable *ROBERT BOYLE*.

An Advertisement of the Publisher's.

*T*Is manifest enough by the beginning of the following *Paper*, that 'twas not intended to come abroad alone, as indeed it was but a part of some *Writings* about *Cold*, defign'd to inlarge the *History* of that *Quality*. But yet the *Author* forbore, by altering it, to accommodate it to the *Papers* wherewith it now comes forth; because in this very form it was by him (being to take a journey) left feal'd up with the *Learned Secretary* of the *Royal Society* in *Febr.* in the year 166⅔; fince when it did, till lately, continue in thofe fafe hands; the *Author* having no need to make ufe of it. Which *Circumstances* are now mention'd to keep the *Reader* from won-dring, that the *Author* fpeaks of the *Production* he made of *Cold* by the *Conflict* of two *Liquors*, as a *New Experiment* and *Phænomenon*; though now two or three years ago, the learned *Sylvius*, as he is inform'd, takes notice in one of his *Books*, of a way of producing *Cold* by a *Mixture* of *Spirit* of *Vitriol*, and another *Saline Spirit*. But befides that the *Author's* way is differing enough from *Sylvius* his, 'tis apparent by the time when his *Experiment* was left with *Mr.* Oldenburg, who is ready to bear witnefs to what is here faid, that he had made it at leaft fome years before the other, to which he was perfectly a ftranger. nor hath yet ever tryed it, came abroad. Nor fhould he eafily have look'd for the *Production* of *Cold* by the mixture of the *Acid Spirit* of *Vitriol* with every *Volatile Spirit*; be-caufe he found that the *Oyl* of *Vitriol* (as that *Acid Liquor* is commonly, but abufively, called) would by its conflict with *Uri-nous Spirits*, produce not *Cold* but *Heat*. Whether the *Care* and *Cautioufnefs*, with which he made the *Experiment* now to be fubjoyned, may give the *Diffident* and *Curious* more fatif-faction than a bare *Affirmation* would do of his having produced *Cold* upon a mixture of contrary *Bodies*, he leaves it to others to judge. And I fhall now only add, that he fome years fince fhew'd fome *Curious Perfons*, that *Cold* may be made to refult from the *Conflict* of *Bodies*, whereof none is a neceffary *Ingre-dient* in the *Experiment*, from which, it may be, I have too long detain'd the *Reader*.

A NEW

EXPERIMENT

ABOUT THE

PRODUCTION of COLD

BY THE

Conflict of Bodies, appearing to make
an Ebullition.

AND now that we are searching after
the Nature of Cold, I am put in mind
that I have sometimes wondred at a
certain Experiment that is so Ano-
malous, and seems so little of kin to
the usual *Phænomena* of Cold, that
though I do not particularly teach the way of making it, because I could not do it without discover-
ing something in Chymistry, that cogent considerations forbid me at present to publish ; yet I cannot
forbear to relate, on this occasion, the matter of
Fact, *both* because it may afford considerable Hints
to sagacious Inquirers, *and* because it seems so little congruous to most Theories of the *Causes* of
Cold, that it may make the Framers of Theories

more wary, and help also to excuse my backwardness to propose *Hypotheses* about Cold in a resolute and confident way.

The Experiment is this : We took three Saline Bodies, each of them purify'd by the Fire ; and whereas there are divers Bodies, that being mingled together acquire a Heat, which neither of them had apart; and whereas it is said by some that there are a few, which being blended together make a mixture somewhat colder than either of themselves, these Salts of ours being put together in due proportion, do upon their mixture produce that, which the Eye judges to be a great Effervescence ; but though the hissing noise be loud, and though the numerous Bubbles suddenly generated will make the matter apt to overflow the Glass, if the one be not capacious, and the other be not put in by little and little ; yet even whilst this seeming Ebullition lasts , the Glass, which one would expect to find very hot, (as usually happens upon the mixture of the Salt of Tartar, and Spirit of Nitre, and upon the contusion of the like Saline Bodies dispos'd to produce together such Efflorescencies) instead of growing hot, does, if it be held in ones hand , feel much cooler than before, and that in a wonderful degree; insomuch that ev'n in Winter the outside of the Glass would quickly be cover'd with great drops of Dew, which after a while would unite, and trickle down by their own weight. And this we could make to last for a great while, by casting in by degrees more and more of one of the Ingredients on the other.   And besides that, this copious Dew on the outside of the Glass, reach'd as high as the mixture within, which argued whence it proceeded ;
be-

befides that, purpofely looking on the bottom of the
Glafs whofe outfide was concave, we found no fuch
drops of Dew there, becaufe the Vapours of the External Air could not in any quantity have accefs to
it ; which fhew'd the Dew, confpicuous elfewhere,
not to come from the tranfudation of the finer parts
of the Mixture through the pores of the Glafs : Befides thefe things, I fay, I remember, that having
fometimes purpofely wip'd off the Dew here and
there with my Handkerchief, the dry parts of the
Glafs would in no long time regain trefh drops of
Dew. And this odd Experiment we did for the
main repeat not only in the prefence of an Indu-
ftrious Chymift, (whofe Tryals unexpectedly gave
us the Rife of the Experiment,) but alfo alone, and
at differing feafons of the year.

I fhall add, that having afterwards, about the
middle of *November*, thought fit to vary a little, and
repeat the Experiment, becaufe I could then make
ufe of a feal'd Weather-glafs, which I had not at
hand when I made the former Tryals ; I took two
deep Glaffes, into the one of which I put a good
quantity of fair Water, and in the other I made
fuch a Mixure as I was lately mentioning ; and
having by a ftring, (to prevent the altering of the
temper of the included Air by the warmth of my
fingers) let down the Weather-glafs into the Water, that the Liquor fhut up in the Inftrument
might be cool'd by the ambient Water ; after it had
ftay'd there a reafonable time, I took it out by the
ftring that was faftened to the upper part of it, and
letting it down into the mixture that was then hiffing, and filling the Veffel that contain d it with
multitudes of fucceffively emerging and haftily va-
nifhing

nifhing bubbles ; I perceiv'd neverthelefs, that the coldnefs of the feemingly effervefcent mixture made the imprifon'd tincted Liquor to fubfide fo low, that from four inches and three quarters (or thereabout) at which height it ftood in the carefully divided ftemm , when the Weather-glafs was taken out of the Water, it fell in a fhort time lower than to one inch and a half.   And becaufe I forefaw that this might feem fcarce credible, efpecially if I fhould relate how fwiftly the imprifon'd Liquor fubfided at the beginning ; I fhall annex, that, for farther fatisfaction of others, I remov'd the Thermometer out of the mixture into the common Water again, where it foon reach'd to fomewhat above four inches and a half ; and not content with that, I put it a fecond time into fome of the frigefactive mixture before it had done foming, in which it fell, as before, fomewhat below an inch and a half, and, prefently after, almoft as low as to an inch.   And having once more put it back into the Glafs that contain'd the Water, the included Liquor re-afcended to above four inches and a half, and this in an excellent feal'd Weather-glafs, whofe ftemme was not in all above ten inches long, with a Ball proportionably big.   And for farther confirmation, I took notice, that, whilft the mixture, by its hiffing noife, and its ftrangely numerous Bubbles, feem'd to be in a ftate of Ebullition, the outfides of the Glafs that contain'd it, were, as far as the mixture reach'd, fo plentifully bedew'd with the condens'd Vapours of the ambient Air, that their weight carried them down in little ftreams which left round about the bottom of the Veffel a pretty quantity of Liquor, that appear'd by its tafte not to have
been

been made by the tranfudation of any of the fharp
and Saline Liquors that were agitated within the
Glafs. There remain'd only one fcruple, which was
fuggefted to me by the remembring of a circumftance, which however, at the making of the fore-
mention'd Tryals, I had not minded, and which
poffibly moft Obfervers would have neglected;
but calling to mind, that the Water, I had made ufe
of to immerfe the Weather-glafs in, was brought
out of a room wherein a Fire was wont (though
not conftantly) to be kept, whereas the Ingredients of the mixture were kept, and put together in
a Chamber, which, though contiguous to the former, had no Chimney in it; I thought fit, for greater circumfpection fake, to let the Water ftand all
night in this laft-mention'd Chamber, that the Ambient Air might have the fame Operation upon it,
as upon thofe Bodies that were to be Ingredients of
the mixture: And then repeating the formerly recited Experiment, though I thought it needlefs to
fpend time to watch, as before I had done, the
greateft difference in Cold betwixt the Water and
the bubbling Mixture; yet by making removes of
the Weather-glafs to and fro, from one Liquor to
another, it fufficiently appear'd, that the greater
coldnefs, remarkable in the mixture, did not before proceed in any confiderable degree (if in any
degree at all) from the Water's not having been
kept in the fame Room with it.

So that by thefe different Tryals it feems manifeft, That the coldnefs of the mixture was not a
Deception of the Senfory, fince it would be difcover'd by the operation, it had, not only upon the
Vapours of the Air on the outfide of the Glafs, but
upon

upon the Thermometer it felf, plac'd in the midft of the mixture, which this laft nam'd circumftance argues to have been cold throughout, and ev'n in its innermoft parts.

And to fhew, how much this ftrange coldnefs depended upon the peculiar Texture of the mixture, or the ftructure of its component Corpufcles, and the peculiar kind of motion that was excited in the tumultuating Particles ; I fhall here fubjoyn a Relation which probably will not appear defpicable ; namely , That in the firft place I took fome of the acid Liquor, the reft of which I had made ufe of to make the mixture, whereof I have been fpeaking ; and put a convenient quantity of fair Water, which had been kept a night or two in the fame room (wherein was no Chimney) with it, that there might be no caufe of fufpicion, that the one had been expos'd to a more or lefs cold Air than the other; and yet thefe two Liquors did fcarce fenfibly differ in coldnefs ; though to difcover whether they did or no , I remov'd from one to another of them a good feal'd Weather-glafs with a very flender ftemm.

And in the next place, I took a convenient quantity of the pure Salt, I had fo often employ'd, and caft it into a Glafs full of Water, which I had kept many hours in the fame room with it, and wherein I had a little before plac'd a feal'd Weather-glafs,that the included Liquor might be brought to the temper of the Ambient Liquor; but upon this Injection, the tincted Liquor of the Thermofcope fubfided fo little, as not to make me look upon this Salt as being it felf extraordinarily Cold, fince other obvicus Salts (that I have at other times

caft

caſt into Water to cool it a little) and ev'n Sea-Salt would (according to my Eſtimate) have refrigerated it as much, if not more. Nor did I obſerve the Glaſs, wherein I was wont to keep ſtore of our Salt, (though I had often occaſion to handle it) diſcloſe to the touch any remarkable degree of Coldneſs; ſo that the coldneſs of our hiſſing mixture could not be attributed to that of either of the Ingredients apart, but was a Quality emerging upon their being blended. Now when I thus made theſe Preparatory Tryals, having afterwards plac'd in the ſame Window (of the Chamber laſt mention'd) a couple of Glaſſes, with common Water in one, and in the other ſome of that mixture, of whoſe frigeſactive power I had very recently made Tryal; I left them to ſtand there together all night, and left alſo ſtanding by them ſuch a ſeal'd Weather-glaſs as I have been mentioning; and the next morning, when all the viſible commotion or agitation of the minute parts of the contrary Salts of the Mixture was quieted, I put the Weather-glaſs firſt into one of thoſe two Liquors, and then into the other, and after remov'd it back into the former again, without perceiving any difference worth minding betwixt the coldneſs of the mixture and that of common Water: And with much the like ſucceſs I repeated the Tryal, after the Water and the other Liquor had ſtood in the ſame room (unfurniſh'd with a Chimney) for near two dayes and nights.

And for farther confirmation, I ſhall add, that having inſtead of the Salt, which I hitherto made uſe of, taken ſome of the Spirit, that was wont to come over together with that Salt, and did ſo abound

with

with it, that a good deal of it lay undiffolved at the bottom of the Liquor ; having, I fay, imploy'd this faline Spirit inftead of the Salt it felf, and having for Tryals fake mix'd with it another Spirit, drawn in my own Laboratory for the purpofe , which to me feem'd as like, as could be made, to that which I had all this while made ufe of ; I found, that the mixture of thefe two Liquors (though it produc'd far fewer Bubbles than I was wont to have) inftead of growing Cold, grew Luke-warm, and quickly impell'd the Liquor in the Weather-glafs, from a little above three inches, to as much above eight ; and yet, befides that this laft Spirit was, as far as I could perceive, and that after the fame manner, drawn from the fame Materials with that I had us'd all this while; the *Smell* and *Tafte*, (which are both of them peculiar and odd enough) concurr'd to manifeft the two Spirits to be of the fame kind.

And, for farther proof, I fhall add, that to fatisfie my felf the more fully, I took a parcel of the fame Liquor, I had lately employ'd with fuccefs in making the Frigorifick Mixture, and yet ev'n this Liquor, which with the dry Salt would queftionlefs have produc'd a Frigefactive Mixture as well as the reft had done, which I had a little before taken out of the fame Viol ; this Liquor (I fay) put to a new portion of the Saline Spirit above-mention'd, though they did not produce minute Bubbles nume-rous enough to make a Fome ; yet the Mixture, in-ftead of growing very cold, grew manifeftly Luke-warm, not only in the Judgment of the Touch, but by its Operation on a good feal'd Weather-glafs, carefully and for a competent while imploy'd to ex-amine

amine the Temper of it. Whereas on the contrary, having purposely kept some of the Frigorifick Spirit by the Fire side, till its temper was so alter'd, that it nimbly enough rarified and impell'd up the Spirit of Wine contain'd in a seal'd Weather-glass, immers'd in it, and having into this Liquor cast some of the Frigorifick Salt, ev'n whilst the Spirit of Wine was rising, and would probably have risen a pretty while longer; this injected Salt, when it began to be dissolv'd, did not only give a check to the rising Liquor, and quickly put a stop to its ascent; but, (as I expected) soon made it subside again, till it fell about three inches or more (which was very much in a short Weather-glass) beneath the Station where the Spirit of Wine had rested, before the Liquor was set by the Fire side; nay, afterwards, I try'd, That a Frigorifick Salt, being well warm'd by the Fire side, did, with an appropriated Liquor, that was also warm'd, produce a coldness manifestly perceivable by the Weather-glass. So that in these cases a Body but moderately cold, nay actually warm, hastily reduc'd one, actually warm, or at least tepid, to a far greater degree of actual coldness than it self had.

These are some of the Experiments I try'd with the Liquors and Salts, of which, upon allowable Considerations, I must now forbear to set down the way of preparing: But that ev'n at present I may not be altogether wanting to the Curious, I devis'd a way of making a *Succedaneum* to this Experiment, which I shall here willingly annex, as that, which though it be much inferiour to what I may one day be at liberty to acquaint the Reader with; yet it will shew the main thing intended, by mani-
feſting,

festing, That Cold may by the mingling of Bodies be produc'd, or increas'd to a degree exceeding that of either of the Bodies that compos'd the Mixture; and this, though at the same time a seeming Effervescence be made by the Bodies, that thus refrigerate each other.

I took then very good Salt of Tartar, and putting to it a convenient quantity of Spirit of Vinegar, I did, whilst the mixture was hissing, (but seem'd to the touch to have refrigerated the Glass that contain'd it,) immerse into it the Ball of a good seal'd Thermoscope, furnish'd with Spirit of Wine. And, though the Weather-glass were not much above a foot long, yet the coldness of this Mixture made the Tincted Liquor descend, hastily enough, two inches and almost a half. And to shew farther, That this Mixture was actually colder than cold Water, removing the Weather-glass out of the Mixture into that Liquor, the tincted Spirit began to re-ascend, and that so nimbly, that in about three minutes (that the Ball of the Thermoscope stay'd under water) the Spirit of Wine had re-ascended about an inch and a half, if not more. And to try whether this coldness of the mixture did proceed from, or depend upon, some Texture of the parts, that was not very permanent, and yet did not quite degenerate, immediately after the Ingredients had ceas'd to work upon one another, I remember, that near an hour after the Ebullition of the Spirit and Salt of Tartar was over, the Thermoscope being remov'd out of the common Water, where it had stood immers'd, into the Mixture, descended about half an inch or more.   For want of Salt of Tartar

tar I could not begin the Experiment anew, and so am not sure it will alwayes succeed uniformly. *

But yet to give my self what farther satisfaction I could, by trying the same Experiment in such a way as might discover, whether or no the *Phænome-non* did not depend upon, or require some peculiar Texture

* *The Author's warinefs was not here a-mifs, he having after-wards found, that this Experiment did not alwayes succeed.*

in the fix'd Salt that had been employ'd; I took some Alcaly (made by diffolving Pot-ashes in fair water, and reducing them by coagulation to a white Salt,) and pouring Spirit of Vinegar to it, I found, That this mixture did not, whilst it hifs'd, grow at all colder, but rather somewhat warmer. And, for farther satisfaction, immerfing into it the Ball of the newly mention'd Weather-glafs, I found, that it afcended in a short time about an Inch, and, being remov'd into the Water, defcended about half an inch; and by making removes of it from one of thefe Liquors into the other two or three times more, I found, That the Spirit of Wine did rife and fall according to what has been newly obferv'd, but its motions upwards and downwards were both lefs than before, and more flow.

F I N I S.

OBSERVATIONS

AND

EXPERIMENTS

ABOUT THE

SALTNESS of the SEA.

By the Honourable *ROBERT BOYLE*.

ADVERTISEMENT

To the following Obſervations (which may alſo ſerve for many Hiſtorical paſſages in the Author's other Writings.)

WHereas the Author does frequently make uſe of the Relations of profeſſed Seamen and other Navigators, and of Obſervations made ſome in the Eaſt, and ſome in the Weſt-Indies, it will be fit to ad-vertize the Reader, that he has been very wary in ad-mitting the informations that he imployes ; being for-ward enough to rejeſt, as he has often done, ſuch as ma-ny others would gladly have received : But notwith-ſtanding his wonted rejeſtion of the particulars he ſaw cauſe to disbelieve, 'twas eaſie for him to be well fur-niſhed with ſuch relations as he makes uſe of ; ſcarce any Writer of Philoſophical things having had ſuch op-portunities of receiving ſuch Authentick Informations from Sea Captains, Pilots, Planters, and other Tra-vellers to remote parts, as were afforded him by the ad-vantage he had in being many years a member of the Council appointed by the King of Great Britain to manage the buſineſs of all the Engliſh Colonies in the Iſles and Continent of America, and of being for two or three years one of that Court of Committees (as they call it) that has the ſuperintending of all the affairs of the juſtly famous Eaſt-Indian Company of England.

OBSERVATIONS

AND

EXPERIMENTS

ABOUT THE

SALTNESS of the SEA.

THE FIRST SECTION.

Chap. I.

The Cauſe of the Saltneſs of the Sea appears by *Ariſtotle*'s Writings to have buſied the Curioſity of Naturaliſts before his time ; ſince which, his Authority, perhaps much more than his Reaſons, did for divers Ages make the Schools and the generality of Naturaliſts of his Opinion, till towards the end of the laſt Century, and the beginning of ours, ſome Learned Men took the boldneſs to queſtion the common Opinion ; ſince when the Controverſie has been kept on foot, and, for ought I know, will be ſo, as long as 'tis argued on both ſides but by Dialectical Arguments, which may be probable on

G 3 both

both ſides, but are not convincing on either. Wherefore I ſhall here briefly deliver ſome particulars
about the Saltneſs of the Sea, obtained by my own
tryals, where I was able; and where I was not,
by the beſt Relations I could procure, eſpecially
from Navigators. ;

First then, Whereas the Peripateticks do, after
their Maſter *Ariſtotle*, derive the Saltneſs of the Sea
from the Aduſtion of the Water by the Sun-beams,
it has not been found that I know of, that where no
Salt or Saline Body has been diſſolved in, or extract-
ed by Water expos'd to the Sun or other Heat,
there has been any ſuch Saltneſs produc'd in it, as
to juſtifie the *Ariſtotelian* Opinion.   This may be
gather'd, as to the Operation of the Sun, from the
many Lakes and Ponds of freſh Water to be met
with, even in hot Countryes, where they lye expoſed to the Action of the Sun.   And as for other
Heats, having out of Curioſity diſtill'd off common
Water in large Glaſs Bodies and Heads till all the
Liquor was abſtracted, without finding at the Bottom the two or three thouſandth part, by my gueſs,
of Salt, among a little white earthy ſubſtance that
uſually remained.   And though I had found a leſs
inconſiderable quantity of Salt, which, I doubt not,
may be met with in ſome Waters, I ſhould not
have been apt to conclude it to have been generated
out of the Water by the Action of the Fire, becauſe I have by ſeveral Tryals purpoſely made, and
elſewhere mention'd, found, that in many places,
(and I doubt not but if I had farther tryed, I ſhould
have found the ſame in more) common Water, before ever it be expoſed to the Heat of the Sun or
other Fire, has in it an eaſily diſcoverable Saltneſs
of

of the nature of common Salt, or Sea-Salt; which two I am not here follicitous to diftinguifh, becaufe of the affinity of their Natures, and that in moft places the Salt eaten at Tables, is but Sea-Salt freed from its Earthy and other Heterogeneities, the ab-fence of which makes it more white than Sea-Salt is wont to be with us. *Thefe laft Words* I add, be-caufe credible Navigators have inform'd me, that in fome Countryes Sea-Salt without any preparation coagulates very white; of which Salt I have had, (from divers parts) and us'd fome parcels.

But fome of the Champions of *Ariftotle's* Opini-on are fo bold as to alledge Experience for it, vouching the Teftimony of *Scaliger* to prove, that the Sea taftes falter at the top than at the bottom, where the Water is affirmed to be frefh. But as for the authority of *Scaliger*, though I take him to be an acute Writer, yet, I confefs that, for reafons elfewhere given, I do not allow it that Veneration which I find given it by very Learned Men, nor am I overprone, even as to matters of Fact, to acqui-efce in what he tells us, when he neither fignifies that he delivers things upon his own Experience, or declares from what credible Information from others he received them.

'Tis true, that having often obferved, that Sea-Salt diffolv'd in Water, is upon the recefs of the fuperfluous Liquor, wont to begin its concretion, not as moft other Salts do, at either the Lateral or Lower parts of the Veffel, but at the top of the Water, I will not think it impoffible, that fometimes in very hot Climates or Weather, the Sea may tafte more falt at the top, than at fome diftance beneath it. But confidering how great a proportion of the

Salt

Salt common. Water is wont to be impregnated with
before it suffers Saline Concretions to begin, and
how far short of that proportion the Salt contained
in the Sea Water is wont to be, infomuch that about
Holland, a *Dutch* Geographer or two have not
found it to amount to the proportion of one to forty,
and I in *England* found it to be no more than I shall

hereafter specifie; it seems not likely
See the third   that *Scaliger's* Observation was well
Section, to-   made, and it muft be very unlikely
wards the lat-   that it should generally hold, if the
ter end.   Saltnefs of the Superficial parts of
the Sea be compared with that of the lower parts
of it.

And yet I do not build my Opinion wholly upon
this Argument of fome Modern Philofophers, That
Salt being a heavier body than Water, muft necefla-
rily communicate moft Saltnefs to the loweft parts.
For though this Argument be a probable one,
yet Water being a fluid body, the reftlefs agitation
of whofe Corpufcles makes them and the Corpufcles they carry with them perpetually shift places,
whereby the fame parts come to be fometimes at
the Top, and fometimes at the Bottom. This con-
fideration, together with what was lately noted of
the peculiar Difpofition of Diffolved Sea Salt, to begin its Coagulation upon the furface of the Water,
may make the Argument we are confidering fufpected not to be fo cogent, as at firft fight one may
think it. Which fufpicion I might fomewhat countenance by fubjoyning, that in divers Metals, and
other tinéted Solutions, I have not ufually obferv'd the upper part of the Liquor to be manifeftly deeper coloured than the lower; though between

tween Metalline Bodies and their *Menftruums,* he
difproportion of fpecifick gravity does ufually much
exceed that which I have met with, between Sea-
Salt and Common Water.

CHAP. II.

'T is urg'd out of *Linfcotten* by a Learned Modern Writer, that wanting frefh Water near
Goa (the Metropolis of the *Portugals* in the *Eaft-
Indies*) they make their Slaves fetch it, by diving,
from the bottom of the Sea, which feems a clear
evincement of the Peripatetick opinion. But in this
Obfervation I cannot acquiefce, for two Reafons :
The one, becaufe that though what is alledged as
matter of Fact were ftrictly true, yet fo general a
conclufion could not be fafely drawn from that particular inftance, fince in other parts of the Sea the
contrary has been found by Experience, as I fhall
fhew ere long. And other reafons than thofe given
by the Peripateticks may be rendred of what happens at *Goa,* which reafons may extend to the like
cafes, if elfewhere they fhall happen to be met
with. For it may very well be, that Springs of
frefh Water may arife in fome parts of the furface
of the Earth, that are cover'd with the Sea, as they
do in innumerable Vallies and other places of the
Terreftrial Surface that is not fo covered. Not to
mention thofe Springs that appear in divers places
upon a low Ebb, cover'd with the Sea during the
Flood. The Curious *Hungarian*   *De Admirandis* Hun-
Governour that gives us an ac-   *gariæ Aquis.*

count

count of the wonderful Waters that ennoble his Countrey, relates, that in the River *Vagus* that runs by the fortrefs *Galgotium*, the Veins of hot Water fpring up in the bottom of the River it felf.

Neque in Ripa tantùm, fayes he, *eruun-*
Pag. 65. *tur calidæ, fed etiam intra amnem, fi fundum ejus pedibus fuffodias; calent autem immodicè*, &c. Nay, I have been affur'd by more than our Learned Eye-witnefs, that there is a place upon the *Neapolitan* Coaft, where they (and I think a Writer or two of thofe parts) obferv'd the Water to fpring up hot beneath the Surface of the Sea, infomuch that one of my Relators thrufting in his hand and arm fomewhat deeper than was convenient, found there an offenfive degree of Heat.

Befides, (which is my fecond conjecture) as to the particular cafe of *Goa*, I had the curiofity to enquire of a great Traveller, and a man of Letters, that liv'd in that City and the neighbouring places, and gave me a pertinent account of them, and efpecially of that place whence the frefh water is fetch'd by the Divers, which his Curiofity led him to vifit, and take fpecial notice of; but I found by him, that the Divers do not now think it needful to fetch their frefh water fo low as from the bottom of the Sea, and that by the little depth, whence his and other mens curiofity caus'd it to be taken up, he judg'd it did not fo much come from any frefh water Springs rifing at the bottom of the Sea, as from a fmall River (whofe name I do not remember) that not far from thence runs into the Sea, with fuch a juncture of circumftances, that at the mention'd places, the frefh water does yet keep it felf
tole-

tolerably diftinct, and is not yet fo far made brackifh, as not to continue potable, though not very
good. Which conjecture of his I could make
probable, by what I have had from eminent and
obferving men among our own Navigators, touching
the fliding of Waters one over another, in fome
parts of the Sea, efpecially near the mouths of Rivers. But the difcuffion of this matter, and the
particulars of the Account given me of the fcitua-
tion of the place where Water is div'd for near
Goa, would require more words than they would
in this place deferve, unlefs the point under debate were more important to our prefent purpofe.

I might here pretend to a clear demonftration
by experience of the contrary of what *Scaliger* delivers, by vouching the teftimony of the Learned *Pa-*
tricius, who affirms, that being upon the Sea which
takes its denomination from the Ifland of *Crete*
(now *Candia*,) he did, in the company of a *Vene-*
tian Magiftrate, *Moccenigo*, let down a veffel (fur-
nifh'd with a weight to fink it) to the bottom of
the Sea, where, by the help of a contrivance, it was
unftopp'd, and fill'd with Water there, which being drawn up, was found to be not frefh but Salt.
This Experiment, I fay, I could oppofe as a Demonftration againft *Scaliger* ; but though it be a very
probable Argument, and more confiderable than any
I have feen brought by the *Peripateticks* for their
Opinion, yet I confefs it would be more fatisfactory
to me, if it would not permit me to fufpect, that in
the drawing up of the Veffel through the Salt water,
though there had been Frefh water taken in at the
bottom, the tafte may have been alter'd by the
fub-

ſubingreſſion of Salt water, which being bulk for
bulk heavier than Freſh, would by its ponderouſneſs endeavour to ſink into the aſcending Veſſel,
and thereby more eaſily expell part of the Freſh
water, and mingle with the reſt. Wherefore I ſhall
confirm the Saltneſs of the Sea at the bottom by ſome
Obſervations, that are not liable to the ſame Objections as that of *Patricius*.

The firſt is that of the Perſon, whom I elſewhere
mention, to be able by help of an Engine to ſtay a
conſiderable time at the bottom of the Sea ; for of
him I learn'd, among other things that I deſir'd to
be inform'd of touching that place, that he found
the Water to have as Salt a taſte there as at the
top.

The next Obſervation I obtain'd by means of a
great Traveller into the *Eaſt* and *Weſt Indies*, who
having had the curioſity to viſit the famous Pearl-
fiſhing at *Manar*, near the great *Cape* of *Comori*,
anſwer'd me, that he had the ſame curioſity that I
expreſs'd to learn of the Divers, whether they
found the Water Salt at the bottom of the Sea
whence they fetch their Pearl-fiſhes ? and that he
was aſſur'd by them that it was ſo : And the ſame
perſon being asked by me about the Saltneſs of the
Sea in a certain place under the *Torrid Zone*,
which the relation of a Traveller inclin'd me to
think to abound extraordinarily with Salt, affirm'd
to me, that not only the Divers aſſur'd him, that
the Sea was there exceeding Salt at the bottom,
but brought up ſeveral hard lumps of Salt from
thence, whereof the Fiſhermen and others were
wont to make uſe to ſeaſon their meat, as he himſelf alſo did ; which yet I may aſcribe not only to
the

the plenty of Salt already diſſolv'd in the Water, but to the greater indiſpoſition, that ſome ſorts of Salts, whereof this may be one, have, to be diſſolv'd in that Liquor.

To theſe I ſhall add this third Obſervation: Meeting with an inquiſitive Engineer, that had frequented the Sea, and had ſeveral opportunities to make Obſervations of other kinds in deep Waters, I deſir'd him that he would take along with him a certain Copper Veſſel of mine, furniſh'd with two Valves opening upwards, and let it down for me the next time he went to Sea; on which occaſion he told me, that (if I pleaſed) I might ſave my ſelf the trouble of the intended tryal, for, with a Tin Veſſel very little differing from that I deſcribed unto him, he had had the curioſity near the Straight of *Gibraltar*'s mouth, (where he had occaſion to ſtay a good while) to fetch up Sea-water from the depth of about forty fathom, and found it to be as ſalt in taſte as the Water near the Surface.

Theſe Obſervations may ſuffice to ſhew, that the Sea is Salt at the bottom, in thoſe places where they were made; but yet I thought it was not fit for me to acquieſce in them, but rather endeavour to ſatisfie my ſelf, by the beſt tryal I could procure to be made with my Copper Veſſel, (as more ſtrong and fit than a Tinn one,) what Saltneſs is to be found in the Water at the bottom of our Seas, not only becauſe it may more concern us to know that, but chiefly becauſe, though I deny not, that in the fore-going Obſervations the taſte may ſufficiently prove that the Sea is Salt at the bottom as well as the top, yet I thought the

taſte,

tafte, by reafon of the predifpofitions and other unheeded affections 'tis liable unto, no certain way to judge whether the top and the bottom be as Salt one as the other. Wherefore I thought it would be more fatisfactory to examine the Sea-water by *weight* than by *tafte*, and in order thereunto, having deliver'd the above-mention'd Inftrument to the Engineer I lately fpake of, when he was going to Sea, he fent me, together with it, a couple of Bottles of Sea-water, taken up, the one at the top, and the other at the bottom, at fifteen fathoms deep. The colour and fmell of thefe two Waters were fomewhat differing; but when I examin'd them Hydroftatically, by weighing a roul of Brimftone firft in one, and then in the other, I fcarce found any fenfible difference at all in their fpecifick gravities. So that if the degree of the Saltnefs of Sea-water may be fafely determined by its greater or leffer weight, then fo far forth as this fingle Experiment inform'd me, the Saltnefs is equal at the top and bottom of the Sea : I faid, *if the degree*, &c. becaufe of what I fhall hereafter take notice of about Salts of lefs fpecifick gravity than Sea-Salt.

CHAP.

CHAP. III.

IT follows now that I make out, what I formerly intimated, That though it were granted, that near *Goa*, and perhaps in fome other places, the Divers may have found the Water frefh at the bottom of the Sea, it would not therefore neceffarily follow, that the Sea-water, generally fpeaking, is Frefh at the bottom; for the Obfervations lately mentioned fufficiently manifeft the contrary : And as to thofe very few places (if really there have been any) where the Sea-water has been found Frefh at the very bottom, I think one may afcribe the tafte of the Water to the bubbling up of Springs of Frefh Water, at, or near enough to, thofe very places. I know this may appear a Paradox, fince it may feem altogether unlikely, that fo fmall a ftream of Water as can be afforded by a Spring, fhould be able to force its way up in fpite of the refiftance of fo vaft a weight as that of the fuperincumbent Sea-water, efpecially fince this Liquor by reafon of its Saltnefs is heavier *in fpecie* than Frefh Water.

But this Objection needs not oblige me to forfake my conjecture; for whatever moft men believe, and even Learned men have taught, to the contrary, it matters not how great the quantity of Liquor be, which is laterally higher than the lower Orifice of the Pipe or Channel that gives paffage to the Liquor that is to be impell'd up into it; provided the upper furface of the Liquor in the Channel or Pipe have a fufficient perpendicular height in reference to that

of

of the ſtagnant Water; for no more of all this fluid will hinder its aſcent, than the weight of ſuch a Pillar of the ſaid fluid as is directly ſuperincumbent on it.. Stevinus and I have by differing wayes particularly proved, that, according to the Laws of the true Hydroſta-ticks, the prevalency of two Liquors that preſs againſt each other, is not to be determined accor-ding to the Quantity of them, but to be adjudg'd to that which exceeds the other in (perpendicu-lar) height; ſo that conſidering the Channel wherein a Spring runs into the Sea, as a long and inverted Syphon, if that part of the either neigh-bouring or more diſtant ſhore, whence the Spring or River takes its courſe, be a neighbouring Hill, or Rock, or any other place conſiderably higher, than that part of the bottom of the Sea (or of the ſhore cover'd with the ſurface of the Sea) at which the Channel, which conveyes Freſh water, termi-nates, that Liquor will iſſue out in ſpite of the re-ſiſtance of the Ocean.

V. Stevinum prop.10. l. 4. Statices. And See the Author's *Hy-droſtatical Paradox-es.*

To illuſtrate at once and prove this Paradox, I thought upon the following Experiment. I took a Veſſel of a convenient depth, and a Syphon of a proportionable length, both of them of Glaſs, that their tranſparency might permit us to ſee all that paſſed within them.. Into the larger Veſſel we put a quantity of Sea-water, and into the longer leg of the Syphon, which had been for that purpoſe in-verted, we poured a convenient quantity of Freſh water, which we kept from running out at the ſhort-er leg, by ſtopping the Orifice of the longer with the

the thumb or finger : Then this Syphon being ſo plac'd in the greater Veſſel, that the Orifice of the ſhorter leg was a great deal beneath the Surface of the Salt water, and the Superficies of the Freſh water in the longer leg was a pretty deal higher than that of the ſurrounding Salt water, we unſtopped the orifice of the upper leg, whereby the water in the Syphon tending to reduce it ſelf to an *Æquili-brium* (or equality of height) in both legs, the water in the upper leg being much higher and hea-vier than that in the other, did, by ſubſiding, drive away the Water in the ſhorter leg, and make it ſpring out at the orifice of the ſhorter leg, in ſpite of the breadth and ſpecifick gravity of the Salt water. And this impelling upwards of the Freſh water laſt-ed as long as the ſurface of that water in the longer leg retained its due height above that of the ſur-rounding Sea water ; which circumſtance I expreſly mention, becauſe there being a difference amount-ing to between a fortieth and fiftieth part betwixt the ſpecifick gravity of our Sea water and common Freſh water, by reaſon of the Salt , which makes the former the heavier, the Freſh water in the long-er leg of the Syphon ought to be between a fortieth and fiftieth part higher than the ſurface of the Sea-water, to maintain the *Æquilibrium* betwixt theſe two Liquors.

To make the fore-mentioned Experiment the more viſible, I thought fit to perform it with Freſh Water ting'd with Braſil or Logwood; but that it might not be objected, that thereby the ſpecifick gravity of the Liquor would be alter'd or in-creas'd, I afterwards choſe to make it with Claret Wine, which being a Liquor lighter than Common

Water, and of a conſpicuous colour, is very convenient for our purpoſe.

And when I made this tryal, by placing the
Orifice of the ſhorter leg at a convenient diſtance
below the ſurface of the Sea-water, 'twas not un-
pleaſant to obſerve, how upon the removal of the
Finger that ſtopp'd the Orifice of the longer leg,
the quick deſcent of the Wine contain'd in that
leg, impell'd the colour'd Liquor in the ſhorter
leg, and made it ſpring up, at its Orifice, into the
incumbent Sea-water, in the form of little red
clouds, and ſometimes of very ſlender Streams.
And as this ſhorter leg of the Syphon was rais'd
more and more towards the ſurface of the Water,
ſo there iſſued out more and more Wine at the
Orifice of it; the Liquor in the longer leg proportionably ſubſiding, but yet continuing manifeſtly higher than the ſurface of the Salt Water,
than which it was *in ſpecie* much lighter.

¶ But here I muſt give an Advertiſement
to prevent a miſtake; for if the Syphon be not
exceeding ſlender, after the Wine in the longer
leg is fallen down to it's due ſtation, a heedful
Obſerver may perceive after a while, that though
the Syphon be kept in the ſame place, there
will iſſue out of the ſhorter leg a little red
ſtream, which proceeds not from the former im-
pulſe of the Wine in the longer leg, but from the
ingreſs of the Sea-water, which being much heavier *in ſpecie* than Wine, ſinks into the Cavity of the Syphon, and as it comes in on one ſide,
thruſts up as much Wine on the other ſide of
the ſame Cavity. But the red Liquor that aſ-
cends on this account may be diſcern'd to do ſo,

by

by its rifing more flowly, and after another manner than that which is impell'd up by the fud-
den fall of the tall Cylinder of Wine in the longer leg.

THE SECOND SECTION.

CHAP. I.

AS to the *Caufe* of the Saltnefs of the Sea, I
therein agree with the Learned *Gaffendus,*
and fome other Modern Writers, That the Sea
derives its Saltnefs from the Salt that is diffolved
in it : But I take that Saltnefs to be fupplied, not
only from Rocks, and other Maffes of Salt, which
at the beginning were, or in fome places may yet
be, found either at the bottom of the Sea, or at
the fides, where the Water can reach them , but alfo (to fay nothing here of what may perhaps be
contributed by fubterraneal Steams) from the Salt,
which the Rains, Rivers, and other Waters diffolve in their paffage through divers parts of the
Earth, and at length carry along with them into
the Sea. For not only 'tis manifeft enough, that
feveral Countryes afford divers falt Springs, and
other running Waters, that at length terminate
their Courfe in the Sea, but I have fometimes fufpected, that very frequently the Earth it felf is
impregnated with Corpufcles, or at leaft, Rudi-
ments of common Salt, though no fuch thing be vulgarly taken notice of. Which fufpicion may be

con-

confirm'd (to omit what I have elsewhere deliver'd on another occasion) partly by the Observation of some eminent Chymists, who affirm themselves to have found a not inconsiderable quantity
of exceeding Saline Liquor upon the evaporation
of large quantities of *some* Waters, (for in some
others I could not find it,) and principally by the
quantity of common Salt that is usually found in the
refining of Saltpeter ; though that be a Salt, which
Sir *Francis Bacon*, and other experienc'd Writers
teach, that almost every fat Earth kept from the
Sun and Rain, and from spending it self in Vegetation, will afford.

But having on another occasion sufficiently shewed, that the Earth does abound
In a Tract of Sub- with common Salt in many more
terraneal Menstru- places than are wont to be taken
ums. notice of; and that 'tis probable,
that by maturation, or otherwise, Salt may daily
grow in the Earth, it will not be necessary to add
in this place any thing to what I have said already
to prove, that our Common Terrestrial Salt being
dissolved, may suffice to make the Sea-water brack-
ish; and the rather, if we call to mind what has been
formerly said about the possibility of Springs rising
beneath the surface of the Sea, and of Lumps of
Salt that were taken up by *Divers*, undissolved, at
the Bottom of the Sea ; the Ocean may receive sup-
plies of Salt from Rocks and Springs latent in its
own Bosome, and unseen even by Philosophers.
And this may be one Reason, I conceive, (for I de-
ry not but that there may be others, as the very
unequal heat of the Sun, &c.) why some Seas are
so much Salter than others, or at least, why in some
pla-

places the Sea-water may be much Salter than in others.

And as we have feen, That our common Terre-ftrial Salt may be copioufly enough communicated to the Sea, to impregnate it with as much Saltnefs as we obferve it to have ; fo I do not fee, that the diffe-rence between that Salt and Sea-falt is fo great, but that it may well be fuppos'd to be derived from thofe Changes that the Terreftrial Salt may be liable to, when it comes into the Sea. For that the Marine Salt and the Terreftrial do very well agree in the main things, may be argued from the refemblance both in fhape, tafte, &c. that may be obferved be-tween the grains that will be produced, if we expofe each of them in a diftinct Glafs to fuch a heat, as may flowly carry off the fuperfluous Moifture, and fuffer them to coagulate into Cubical or almoft Cu-bical Graines : And the leffer differences that may be met with between thefe two Salts, may well enough be fuppos'd producible by the plenty of Nitrous, Urinous, and other Saline, to which, in fome places, may be added, Bituminous bodies, that by Land-floods and otherwife are from time to time carried into the Sea, and by feveral things that hap-pen to it there, efpecially by the various agitation 'tis put into by Tides, Winds, Currents, &c. and (which I would by no means omit) by its being in vaft quantities expos'd to the Sun and Air.

CHAP.

CHAP. II.

WE may juftly be the more careful to determine, whether the Saltnefs of the Sea-
water proceed from Common Salt diffolved in it,
becaufe if it appeared to be fo, we might the more
hopefully attempt to obtain by diftillation Sweet
water from Sea-water; fince, if this Liquor be made
by the bare diffolution of Common Salt in the
other, 'tis probable, that a feparation may be made
of them, by fuch a heat as will eafily raife the Aqueous parts of Sea-water, without raifing the Saline,
whofe Diftillation requires a vehement Heat, as
Chymifts well know to their coft. And fuch a me-
thod of feparating Frefh water from that which was
Salt, would make our Doctrine of ufe, and be very
beneficial to Navigation, and confequently to Man-
kind. For in long Voyages, 'tis but too common
for the makers of them, to be liable to hazards and
inconveniencies, for want of Frefh and Sweet water,
whereby they are fometimes forced to drink corrupt
brackifh Water, which gives them divers Difeafes,
as particularly the Scurvy, and, the ufual effect of
drinking Salt water, the Dropfie. And Sea-men
are wont to receive fo many other incommodities
by the want of Frefh water, that, to prevent or fupply it, they are oftentimes forced to change their
courfe, and fail fome hundreds of miles to a Coaft,
not only out of their way, but unfafe in it felf, and
perhaps more dangerous, by being infefted by Py-
rats, or in the hands of Enemies or Savage people;
by which meanes they often lofe the benefit of
their

their *Monfouns*, and much more eafily other Winds,
and frequently their Voyage. And thefe are inconveniencies, which might be in good meafure
prevented, if potable, and at leaft tolerably wholfome Water, could be obtain'd by Diftillation, in
the midft of the Sea it felf, to ferve the Sea-men till
they could be fupplied with naturally Frefh water.
To make fome tryals of this , I remember I took
fome *Englifh* Sea-water, whence I was able to fe-
parate betwixt a thirtieth and fortieth part of dry
Salt, and having diftilled it in a glafs head and body, with a moderate fire, till a confiderable portion
of it was drawn over , we could not difcern any
Saltnefs in it by the tafte; and befides that I found
it fpecifically lighter than fuch Water as is daily
drunk by Perfons of Quality at *London* , I expos'd
it to a more Chymical Examen, and did not by that
find any thing of Sea Salt in it, though I have
at feveral times, by the fame way, manifeftly difco-
vered a Saltnefs in in-land Waters, that are drunk
obvioufly for fweet Waters. If I would have employed a ftronger Heat, and Veffels larger and
lower, or otherwife better contriv'd for copious
Diftillation, I might in a fhorter time have obtain'd
much more diftill'd Water ; but whether fuch Liquors will be altogether fo wholfome , Experience
muft determine. Yet that Sea-water diftill'd
even in no very artificial way, may be fo far wholfome, as not in hafte to be fenfibly noxious, but at a
pinch ufeful, at leaft for a while, may be gathered
from (what occurrs to me fince the writing of the
laft Paper) the Teftimony of that famous Naviga-
tor, Sir *R. Hawkins*, who commanded a Fleet in
the *Indies* for Queen *Elizabeth.* For he, in the Ju-

dicious

dicious Account he gave the World of his Voyage, wherein they were diſtreſſed, even in the Admirals ſhip, for want of Freſh Water, has this memo-

In Lib. 7. page 1378. of rable paſſage (as I find it *ver-*
Purchaſe; out of Sir *batim* in our diligent *Pur-*
R. Hawkins his Voyage. *chaſe.*)

Although our freſh water had fail'd us many dayes (before we ſaw the ſhore) by reaſon of our long Navigation without touching any Land, and the exceſſive drinking of the Sick and Diſeaſed (which could not be excuſed;) yet with an invention I had in my Ship, I eaſily drew out of the Water of the Sea ſufficient quantity of Freſh water, to ſuſtain my people, with little expence of fewel; for with four billets I ſtill'd a hogſhead of Water, and therewith dreſſed meat for the Sick and Whole. The Water ſo diſtill'd we found to be wholſome and nouriſhing.

And becauſe the potableneſs of Sea-water may concern the Healths and Lives of men, I ſhall here add, to what I elſewhere deliver about my wayes of examining, whether other waters participate of Salt, two or three Obſervations I made upon thoſe few diſtill'd Liquors, I had occaſion to draw from Sea-water. Having then upon ſome of the diſtill'd Liquor dropt a little oyl of Tartar *per deliquium,* I perceiv'd no clouds at all, or precipitation to be made, whereas a ſmall proportion of that Liquor being dropt into the undiſtill'd Sea-water it ſelf, it would preſently trouble and make it opacous, and, though but ſlowly, ſtrike down a conſiderable deal of a whitiſh ſubſtance (which, of what nature it is, I need not here declare ;) I found alſo, that a very ſmall proportion of an Urinous Spirit, ſuch as that of *Sal Armoniac,* would produce a whitiſh and curl-

ed

ed fubftance (but not a near fo copious one as the
other Liquor *)* in Sea-water, not yet expos'd to
Diftillation, but not in the Liquor drawn from it :
which argued, that there were but few or no faline
particles of Sea-falt afcended with the Water : For
elfe thefe Alcalizate and Urinous Salts would in all
likelihood have found them out, and had a vifible
operation on them.　And I farther remember, that
when the Diftillation was made in Glafs Veffels,
with an eafie Fire, not only the firft running,
but the Liquor that came over afterward's, was not
perceiv'd to be brackifh, but good and potable.
To which agrees very well, that by a Hydroftatical Tryal I found our diftill'd Sea-water to be
lighter *in fpecie* than common Conduit Water,
though it exceeded *that* in fpecifick Levity, lefs
than 'twas furpaffed in the fame quality by diftill'd
Rain-water.

But to return to the Subject whence we have
fomewhat, but, I hope, not ufelefly, digrefs'd ; I
know it may be objected, that if the Terreftrial Salts carried by Springs, Rivers, and Land-
floods into the Sea, were the caufe of its faline
Tafte, thofe Waters themfelves muft be made Salt
by it, before they arrive at the Sea.　But befides,
that this Objection will not reach the Springs and
Rivers of Salt water, that in feveral places, either
immediately or mediately, difcharge themfelves into
the Sea ; it might conclude againft him that fhould
affirm this imported Saltnefs to be the only caufe of
that of the Sea : But it will not be of force againft
me, who take it to be only a partial caufe, that by
its acceffion contributes to the degree of Saltnefs
we obferve in the Sea, where this imported Salt
may

may joyn it ſelf with the Salt it finds there already, and being detained by it, contribute to the briny-neſs of the Water.

If it be urg'd, that from hence it will follow, that the Sea from time to time increaſes in Saltneſs, I may ſuſpend my anſwer till it appear by competent obſervation that it does not ; which, I think, men have not yet made tryals that may warrant them to aſſert. And if the matter of fact were certain, I think 'twere poſſible to give a farther anſwer, and ſhew probable wayes, how ſo ſmall an acceſſion of Salt may be diſpers'd by nature, and kept from increaſing too much.

CHAP. III.

BUt now 'tis ſeaſonable to conſider, that the taſte of Sea-water is not ſuch a ſimple ſaline taſte , as Spring-water would receive from *Sal Gemm*, or ſome other pure Terreſtrial Salt diſſolved in it , but a bitteriſh taſte, that muſt be derived from ſome peculiar cauſe that Authors are not wont to take notice of. I am not aſſur'd by any Obſervations of my own, that this receſſion from a purely Saline taſte is likely to be of the very ſame kind, and to be equally, or very near equally, met with in all Seas ; (not to add a doubt whether it be at all ſenſible in ſome.) The cauſe both of the bitterneſs and ſaltneſs too of the Sea-water, is ſaid to be af-firmed by Learned Mr. *Lidiat*, to be aduſt and bi-tuminous

tuminous Exhalations aſcending out of the Earth into the Sea. But that there is abundance of actual Salt in the Sea-water, to give it its Saline taſte and ponderouſneſs, the Salt, that the Sun does in many places copiouſly ſeparate from the Saltleſs wateriſh parts, ſufficient·y manifeſts. But as to the bitteriſh taſte, I think it no eaſie matter to give a true account of it, but am prone to aſcribe it *partly* to the operation of ſome Catholick Agents upon that vaſt body of the Ocean, and *partly* to the Alteration that the Salt receives from the mixture of ſome other things, among which *Bitumen* may be one of the principal.

But though I have in another Paper ſhewn, that in ſome places of the Sea there are conſiderable quantities of *Bitumen*, or Bituminous *In the Tract of Subterran.* Menſtruums. matter to be met with; yet I dare not derive the bitterneſs of the Sea only from Bituminous Exhalations, but in good part, at leaſt, in ſome places, from the liquid and other Bitumen, that is imported by Springs and other Waters into the Sea; of which we have an eminent inſtance in that which our *Engliſh* call *Barbadoes* Tar, according to the relation I had of it from an inquiſitive Gentleman, who is one of the chief Planters of the Iſland, and took pleaſure to obſerve this liquid Bitumen to be carried in conſiderable quantities from the Rocks into the Sea; and I think it poſſible enough, that ſome of the Springs that riſe under the ſurface of the Sea, may carry up with them Bituminous matter, which may help to make the Saltneſs of the Sea degenerate, (of which more perhaps elſewhere;) as I not long ſince made mention of Springs, as well of hot

as

as cold water, rising beneath the surface of the Sea. And this minds me to intimate here, that I have suspected, that in some places the Sulphureous Exhalations, and other emissions from the submarine parts of the Earth, may sometimes contribute to change the saline taste of the Sea-water: For I have elsewhere related, how not only Sulphureous Steams, but sometimes Actual Flames have broken through from the lower parts of the Sea to the uppermost; and have sometimes taken pleasure to make by Art a rude imitation of that *Phænomenon.* And partly some Experiments of my own, and partly some other Inducements, have perswaded me, that divers times (for I do not say alwayes) Sea Salt does not obscurely participate of Combustible Sulphur, of which I may speak farther on another occasion. But in regard that the taste of the Sea-water is not in all parts of the Ocean uniform, it may here suffice to take notice in general, that this difference of taste may partly be caus'd by adventitious bodies of several kinds, of which 'tis probable, that in differing places the Sea-water does variously partake. And not to mention here the fragrant smell of Violets, which has by several, and particularly by an Eminent Person, of whom I enquired about it, been observed, in some hot Countries, to proceed from Sea Salt ; I have divers other Inducements to think that it is usually no simple Salt, nor free from mixture. For by more wayes than one, and particularly by cohobating from it its own Spirit, we have obtained a dry Sublimate, which seemed to be no Pure, but a Compounded Body.

And now to come to that which I intimated might be one of the caufes, why the tafte of Sea-water is not the fame with that of Common Salt diffolved in Frefh water; I fhall add, that I have fufpected, that the various motion of the Sea, and its being expofed to the action of the Air and Sun, may contribute to give it a tafte other than Saline; which fufpicion might be confirmed by the Obfervation I elfewhere mention of the Sea Salt, which, by barely being expos'd for many months to the Air, and fome-times perhaps put into a gentle agitation by a dige-ftive Heat, I found to have a very manifeftly differ-ing tafte from the fimple Solution of Sea Salt in Common water.

I might here endeavour the farther confirmation of my Difcourfe, by what I have learned by in-quiry from Navigators, about the manifeftly differ-ing Colours, and other Qualities of the differing parts of the Sea, which feem to argue, that 'tis not every where of fuch a Uniform Subftance as men vulgarly imagined, and that vaft Tracts of it are imbued with ftupendious multitudes of adventitious Corpufcles, which, by feveral wayes diverfifying its parts, keep it from being a fimple Solution of Salt. But of this Subject I have not leifure to dif-courfe here, only becaufe 'tis generally thought, that the Sea-water is, by reafon of the Saltnefs it a-bounds with, uncapable of Putrefaction; I will add, That having kept a pretty quantity of Sea-water, that I had caufed to be purpofely taken up between the *Englifh* and *French* fhores, in a good new rund-let, in a place where the Summer Sun beat freely upon it, it did, in a few weeks, acquire a ftrongly ftinking fmell; though, that the Experiment had

.been

having failed often in the *Indian* and *African* Seas, I enquired of him, whether he had ever in thoſe hot Climats, where the Sea is ſuppoſed to be very Salt, obſerved it to ſtink, for want of Agitation or otherwiſe : To which he anſwer'd, That once being, though it was but in *March*, becalmed, in a place he named to me, for 12 or 14 dayes, the Sea, for want of motion, and by reaſon of the Heat, began to ſtink, inſomuch that, he thinks, if the Calm had continued much longer, the ſtench would have poyſoned him : They were freed from it as ſoon as the Wind began to agitate the Water, and broke the Superficies, which alſo drove away ſtore of the Sea Tortoiſes, and a ſort of Fiſh, whoſe *Engliſh* name I know not, that before lay basking themſelves on the top of the Water.

And to this agrees very well the notable Obſervation, that I ſince met with, of the elſewhere commended Sir *R. Hawkins*, who, among other conſiderable things he takes notice of in his Relations, has this paſſage, to our preſent

Purchaſe's Pilgrims in Sir *R. Hawkins* Obſervations.

purpoſe. *Were it not for the moving of the Sea by the force of Winds, Tides, and Currents, it would corrupt all the world. The Experience I ſaw* Anno 1590, *lying with a Fleet about the Iſlands of* Azores, *almoſt ſix months, the greateſt part of the time we were becalmed ; with which all the Sea became ſo repleniſh'd with ſeveral ſorts of Gellies, and forms of Serpents, Adders and Snakes, as ſeem'd wonderful,*

ſome

*fome green, fome black, fome yellow, fome white, fome
of divers colours, and many of them had life ; and
fome there were a yard and a half, and two yards long,
which had I not feen ; I could hardly have believed.
And hereof are witneffes all the company of the Ships
which were then prefent, fo that hardly a man could
draw a bucket of water clear of fome corruption. In
which Voyage, towards the end thereof, many of every
Ship fell fick of this Difeafe, and began to dye apace,
but that the fpeedy paffage into our Country was a remedy to the crazed, and a prefervative for thofe that
were not touched.*

THE THIRD SECTION.

CHAP. I.

AS for the various *Degrees* of the *Saltnefs* of the
Sea, Authors are wont to be filent of it,
fave that fome Navigators tell us, that they obferved
fome Seas to have a more, and others a lefs Saline
tafte ; which you will eafily believe has not afforded me much fatisfaction. And on the other fide,
my want of opportunity to make Tryals my felf,
will confine me to acquaint you with no more than
the few following Obfervations.

1. To a Learned man that was to fail to places of
differing Latitudes in the *Torrid Zone*, I deliver'd a
Glafs Inftrument, elfewhere defcribed, fitted by the
greater or leffer Emerfion of the upper part, to fhew,

accu-

accurately enough for use, the greater or less specifick Gravity of the Salt Water it was put to swim in. This he put from time to time into the Sea-Water, as he sailed towards the *Indies*, whence he wrote me word, *That he found, by the Glass, the Sea-water to increase in weight, the nearer he came to the Line, till he arrived at a certain degree of Latitude, as he remembers, it was about the thirtieth; after which, the Water seemed to retain the same Specifick Gravity, till he came to the* Barbadoes *or* Jamaica.

2. Another Observation I obtain'd by Inquiry of an Ingenious Person and a Scholar, at his return out of the *East Indies*, who affirm'd to me, that he, and a Gentleman of my acquaintance, took up Bottles full of Sea-water, both under the *Equinoctial*, and also off the *Cape of good Hope*, which lies in about 34 Degrees of Southern Latitude, and found the Waters of these distant parts of the Ocean to be of the same weight. And though it may well be doubted, whether this Observation, being made with ordinary Bottles, were so exact as could be wish'd, yet the Persons being curious, and making it for their own satisfaction; and my Relator having, in both the recited places, fill'd with the Sea-water he took up and weigh'd, having, I say, fill'd the *same* Bottles; *since* this Vessel held two quarts, (which must be above four pounds of Salt-water,) if the disparity of weight had been *considerable*, it would in likelihood have been found, at least manifestly *sensible* in such a weight of Liquor.

3. Inquiring of an observing Person, that had been at *Mosambique*, which is thought to be one of the hottest places in the World, whether he did

not

not there find the Sea to be more than ordinarily
Salt; he anſwered me, that, coming thither in a
great Carack, when he came back from the Town
to the Ship, he obſerv'd near two hands breadth of
the Veſſel to be above the ordinary part, to which
it uſed to ſink; inſomuch that he took notice of it
to the Captain, as fearing that part of the lading
had been by ſtealth carried to the ſhore : But the
Pilot, who had made thirteen or fourteen Voyages to the Indies, aſſur'd him, what he had obſerved about the Ship was not unuſual in that place,
where the taſte it ſelf diſcover'd the Water to be
exceeding Salt.

Nor need we ſcruple to think, that ſome Sea-
Waters may be very much more impregnated
with Salt than ours; for Water will naturally
diſſolve, and retain a far greater proportion of
Salt, than that which is commonly met with in the
Sea. For whereas a thirty fifth, or thirtieth, or at
moſt a twenty fifth part of Salt will make Water
more Saline than is found in many Seas, I am by
a Friend of mine that is Maſter of a Salt-work,
inform'd, that the Water of his Springs afford
him a twelfth part of good White Salt, and that
another Spring not far off, yields no leſs than an
eighth part. To which, (to avoid anticipation)
I ſhall not here add, what I ſhall hereafter have
occaſion to ſay of the fulleſt impregnation of Water
with Common Salt.

[Whilſt I was reviewing theſe Papers, there
came ſeaſonably to my hands a Letter written from
Muſilapatan, on the Gulf of *Bengala* in the *Eaſt*
Indies, by an ingenious Gentleman, Sir *William*
Langhorn, that is intruſted with the care of the *Eng-*

I   *liſh*

liſh Factories in thoſe parts; out of which Letter the following paſſage is *verbatim* tranſcribed.
" I did, in order to your command, cauſe ſome Wa-
" ter to be ſaved under the Line, at our firſt acceſs
" to it, intending, for want of good ſcales and
" weights, (being none to be come at aboard the
" Ship) to have kept it until it could be weighed,
" but by the forgetfulneſs of a ſervant, it was thrown
" away. Off the Cape, in 37 *d.* 00 *m.* Southern la-
" titude, I ſaved ſome again, and through the ſame
" want of weights, was fain to keep it until I came
" to the Line again; and then made the beſt ſhift I
" could for weights, and compar'd it with the Wa-
" ter there, filling the ſame Bottle again to the ſame
" height by a mark, and found it exactly the ſame
" weight. The weight I have taken; but accounting
" this a journey of buſineſs, left thoſe notes, and
" moſt of the like nature, behind me; in my next
" it ſhall be inſerted.]

CHAP. II.

IT remains now, that, according to my promiſe, I ſet down what I obſerved my ſelf concerning the Saltneſs of our Sea between *England* and *France*; not in compariſon with the Saltneſs of other Seas, whoſe Waters I had not to compare with, but as to the proportion of Salt contained in it to the Water. And though one would think it very eaſie to make tryals of this ſort for a perſon not
un-

unacquainted with Hydroftatical practices nor un-
furnifhed with Inftruments, yet, I confefs, that
three or four tryals that I made; not all of them the
fame way, made me find it more difficult than was
imagin'd to arrive at any thing of certainty in this
inquiry.

This you will eafily believe, if I annex the fubftance of fome Experiments, that, I remember, I
made about the gravity of Sea Water; which I had
order'd to be taken up, fome at the depth of about
fifteen Fathom fomewhat near our fhoar; and fome
in another place of the Channel between *England*
and *France*.

The fum of the firft Experiment is this : We
took a Vial, fitted with a long and ftrait neck, purpofely made for fuch tryals, and having counterpois'd it, fill'd it to a certain height with common
Conduit water: We noted the weight of that Liquor;
which being poured out, the Vial was fill'd to the
fame height with Sea Water, taken up at the furface,
and by the difference between the two weights, the
Sea water appear'd to be about a forty fifth part
heavier than the other.

The fecond Tryal (which was for more accuratenefs made Hydroftatically,) I find regifter'd to
this effect : We carefully counterpois'd in
the Scales, formerly made ufe of, a piece of Sulphur in the upper Sea water, formerly mention'd ;
it weigh'd $\mathfrak{Z}\beta$ + 10 $\frac{1}{2}$ *gr.* and being alfo weigh'd
in the Sea water fetch'd from the bottom, gave us the
fame weight $\mathfrak{Z}\beta$ + 10 $\frac{1}{2}$ *gr.* which fhew'd thofe
two Waters to be of the fame Specifick Gravity :
And then to compare this with the gravity of common Water; we weigh'd the fame Sulphur in com-

mon Conduit Water, and found it $\mathfrak{Z}\beta + 15\frac{1}{2} gr$: By which it appear'd, that the Sea-water was but about a fifty third part heavier than this Water: which is such a difference from the proportion found out by the former way of tryal, that I could not well imagine what to attribute it to, unless the Sea-water by long standing in a Vessel, which, though cover'd, was expos'd to the hot Sun, may both have been rarified, and have had some separation made of its Saline or other heavier parts, on which score that portion we took up for our tryal, might appear lighter than else it would have done; or unless, the Experiment having been made in *London*, where great and sudden rains and other accidents will sometimes visibly vary the consistence of common Water, the Liquor, I then employ'd without examining it, might be more ponderous at that time than at another. To which latter suspicion I was the more inclin'd, because, having afterwards weigh'd the same piece of Sulphur by help of the same ballance in distill'd rain water, I found the weight of the former liquor to exceed that of the latter by a good deal less than a thirty fifth part; which seem'd to make it probable, that if the Water, we chanc'd to employ, had been free from all Saline and other heavy particles, the difference formerly mention'd betwixt this Observation and the fore-going would not have been near so great as it was.

The last way I made use of to examine the proportion betwixt Sea-water and Fresh, was Chymical; whereof my Register affords me this account.

A pound (*Haverdupois* weight) of the upper Sea-water, was weigh'd out, and put into a head and body

body to be diftill'd in a digeftive furnace *ad ficci-*
tatem, and the Diftillation being leifurely made, the
bottom of the glafs was almoft cover'd with fair
grains of Salt, fhot into Cubical figures, and more
white than was expected; in the reft of the coagu-
lated matter we took not notice of any determinate
fhape. The Salt being weigh'd amounted to ʒβ,
Haverdupoi, and 10 *gr.* At which rate the proportion of the Salt to the Water will be that of 30
and $\frac{72}{100}$ to one, and fo will amount to near the thirtieth part; which was fo much greater than the former wayes of tryal made us expect, that I know not
whether it may not be worth while to try, whether
fuch a flow abftraction as we employ of the fuperfluous Water, and our doing it in clofe Veffels, may
not have afforded us more Salt than elfe we fhould
have obtain'd.

To this Relation I find this note fubjoyn'd : Su-
fpecting that there may have fomewhat elfe concurr'd to our finding fo great a proportion of Salt,
I fuffer'd that, which had been weighed, to continue
a while in the Scale, and foon perceiv'd, that, according to my conjecture, that fcale began manifeftly
to preponderate, and that confequently fome of the
unexpected weight of Salt may be due to the moifture of the Air, imbib'd after the Salt was taken
out of the Glafs, and laid by to be weighed : Wherefore, caufing it to be very well heated and dried in
a Crucible, we found it to weigh ʒiij. + β. (that
is 210 *gr.*) upon which account, the proportion of
Salt contain'd in the Water was a thirty fixth part,
and fomewhat above half of thofe parts, and to ex-
prefs it in the neareft whole number , a thirty fe-
venth part.

From whence this greater proportion of Salt by Diftillation, than our other Tryals invited us to expect, proceeded, feems not fo eafie to be determined; unlefs it be fuppofed (as I have fometimes fufpected) that the Operation, the Sea-water was expofed to in Diftillation, made fome kind of change in it, other and greater than before-hand one would have look'd for ; and that, though the grains of Salt we gained out of the Sea-water, feem'd to be dry before we weigh'd it, yet the Saline Corpufcles, upon their concreting into Cubes, did fo intercept between them many fmall particles of Water, as not to fuffer them to be driven away by a moderate warmth, and confequently fuch grains of Salt may have upon this account been lefs pure and more ponderous than elfe they would have been. And I might here add, that I fometimes make a certain Artificial Salt, which though being diffolv'd in Water, it will fhoot into Cryftals finely fhaped, and dry enough to be reducible into powder, yet coagulates Water enough with it to make the Water almoft, if not quite, as heavy again as before. And I have been affured by a very Learned Eye-witnefs, that there is a fort of Sea Salt, which they bring to fome parts of *England* from the Coaft of *Spain* or *Portugal*, which being here diffolved, and reduced by Purification and Filtration to a much whiter Salt, will yield by meafure fomewhat above two Bufhels for one. But to fatisfie the fcruples and fufpicions I could fuggeft, would require more tryals than I have now time or opportunity to make. What has been already deliver'd, may give at leaft as fcrupulous an account of the Saltnefs of our *English* Sea-waters, as moft other Experimenters would

have

have thought it needful to give.   And to make a determination with any certainty about the degrees of the Seas Saltnefs *in general*, a great number of Obfervations, made in different Climates and in di-ftant parts of the Ocean, would be neceffary.

CHAP. III.

I Know not whether I may be fo indulgent to my fufpicions as to wifh, that Obfervations were heedfully made, Whether in the fame Sea, and about the fame part of it, the Waters be alwayes equally Salt ? For, though that be taken for granted, yet fince we have no good Obfervations long fince made to filence the fufpicion, one may fufpect, that, at leaft in many places, the Saltnefs of the Sea may conti-nually, though but very flowly , increafe by the ac-ceffion of thofe Saline Corpufcles that are imported by Salt-Springs, and thofe which Rivers and Land-floods do from time to time rob the Earth of. And I fufpect it to be not impoffible, that this or that part of the Sea may be fometimes extraordinarily, and perhaps fuddenly, impregnated with an additio-nal Saltnefs from Saline fteams plentifully afcend-ing into it, from thofe Subterraneal Fires, about which I have made it *In the Tracts of* elfewhere probable, that they may *Subterran. Fires* burn beneath the bottom of the Sea, *and Steams.* and fometimes fend forth copious Exhalations into it. But it may prove the more difficult to difcern.

I 4   this

this adventitious Saltness, unless the taste as well as ballance be employed about it; because the Salt, that produces it, may be of such a Nature as to be much lighter *in specie* than common Sea Salt. And the mention of this leads me to give you here the Advertisement I promised you not long ago.

That though the weight of Sea-water be as good a way as is yet employ'd (and better than some others) to determine what Sea-water does most abound in Salt; and though it be possible, that in our Sea, and perhaps in almost all others, this way be not liable to any considerable uncertainty; yet I think it not impossible, that it may sometimes deceive us, especially in very hot Regions; because I have observed, that there may be Volatile Salts, which, though by reason of their activity they make smart impressions on the tongue, and give the water imbued with them a strong Saline taste, yet they add very little, and much less than one would think, to its Specifick gravity: as I have tryed, by Hydrostatically examining Distill'd liquors, abounding in Volatile and Urinous Salts, some of which I found very little; heavier than Common Water, and consequently nothing near so much heavier as they would have been made, if they had been brought to so sharp a taste, by having nothing but common Sea Salt dissolved in them: So that, if in any particular place, by any other way, or from the Steams of the Earth beneath, (some of which, I elsewhere shew, may be very analogous to those afforded by *Sal Armoniack*) the Sea should be copiously impregnated with such kind of light Salts, the Sea-water may be much more salt to the taste, and yet be very little heavier. For confirmation of which I find among my notes, that

weigh-

weighing a feal'd buble of Glafs, made heavy by an included Metal , firft in Spirit of *Sal Armoniack,* that tafted much ftronger than Sea-water,it weighed ℥iij + 51 ¼ *gr.* and weighing this fame body in fair Water, it weighed but ℥iij + 45 ¾ *gr.* fo that notwithftanding its great Saltnefs, the Spirit was lighter than Common water ; though a good part of that comparative Levity may probably be afcribed to the Liquor wherein the Saline Particles fwarm, which, by Diftillation , was grown more defecated and light than Common, though clean, Water.

But for a farther proof, we took a hard lump of *Sal Armoniack ,* and though we could not weigh it in Water, becaufe that would have diffolv'd part of it, yet by a way (I elfewhere teach) I found, that weighing in the fame Liquor this lump of *Sal Armoniack,* and a lump of good white Sea Salt , (brought me as a Curiofity out of the *Torrid Zone*). the proportion of the latter to a bulk of the Liquor equal to it, was fomething (though exceeding little) above that of two and a quarter to one , and the proportion of *Sal Armoniack* to as much Water as was equal likewife to it , did not above a Centefm exceed that of one and $\frac{7}{10}$ to one ; which falls fo fhort of the other proportion as may juftly feem ftrange , efpecially if it be confidered, that the factitious *Sal Armoniack,* the Chymifts generally ufe, and we employ, confifts in good part of Sea Salt, which abates much of the Comparative Levity it might have,if it were made up only of Urinous and Fuliginous Salts , which were its other ingredients.

It were indifcreet for me to propofe any more fufpicions and tryals fitted to clear tnem , unlefs I

knew thofe I have already mention'd would not pafs
for Extravagancies ; and therefore I fhould here difmifs the Subject of this Tract of the Saltnefs of the
Sea, but that fince I have been difcourfing of the
degrees of it, it will not be impertinent to add,
what is the greateft meafure of Saltnefs that I have
brought Water to, without the help of external
Heat. On this occafion I employed two differing
wayes, the one was by putting into a well-coun-
terpoyfed Vial two Ounces of Common-water,
and then putting into it well dryed and white common Salt, and fhaking them together till the Liquor would, *whilft cold,* diffolve no more : This Liquor, thus glutted with Salt, weighed 1150 grains,
from which two Ounces being deducted, the
overplus of weight, arifing from the diffolved Salt,
amounted to 190 *gr.* fo that a parcel of Salt will
without heat be diffolved in about five times its
weight, or very little more, of common Water.
By which proportion we made fo ftrong a brine,
that divers pieces of Amber, being purpofely let
fall into it, emerged, and floated on it. The other
and better way, yet more tedious, that we made
ufe of, was, to let Sea-Salt run *per deliquium,* (as
the Chymifts fpeak) that is, to fet it in fome moift
place, till it was diffolved by the Aqueous Vapours that fwim in the Air. In this Liquor we
weighed a piece of Sulphur, which we alfo weighed
in Sea-water, wherein, finding it to weigh much
more than in the former Liquor, it appeared that
the Sea-water was *in Specie* much lighter than
the other ; though how much their gravities differ'd, I cannot find among my Notes, nor be inform-
ed by my Memory.

And

And becaufe I have not in any Author met with the proportion of Sea Salt to Water of the fame bulk, nor perceive that Hydroftaticians themfelves have yet attempted any way to inveftigate it, (probably deterr'd by the eafie diffolublenefs of Salt in Water) I fhall here fubjoyn , that by the help of an Expedient I have elfewhere taught, I have examin'd a hard dry lump of Sea-Salt , and found its proportion in weight to common Water of the fame bulk, to be *almoft* as **2** to **1**, (for it exceeded the *ratio* of **1** $\frac{9}{10}$ to **1**.) And, I remember, I found the Specifick Weight of a hard and figur'd lump of *Sal Gemm* (which fort of Salt, I fuppofe to be fomewhat more pure and ponderous than Sea Salt) to be to that of Water (very near) as **2** $\frac{1}{2}$ to **1**.

F I N I S.

THE FOURTH SECTION

Belonging to the Tract formerly Publish'd under the Title,

Relations about the Bottom of the S E A.

By the Honourable *ROBERT BOYLE.*

ADVERTISEMENT

TO THE

READER

THis Section should have been subjoyned to the Relations about the Bottom of the Sea, when that Discourse was printed, together with some other Tracts at Oxford, An. 1671. but was by the Negligence of him, that should have carried it to the Press, severed from the rest of that Tract, and not seasonably deliver'd to the Printer.

THE FOURTH SECTION

Belonging to the Tract intitul'd,

Relations about the Bottom *of the* SEA.

He prefence of the Air is not only fo neceffary to the Life of many forts of Animals, but it hath likewife fo great a ftroke in the growth of Vegetables, efpecially of the larger forts, that, after what I had experimented about thefe matters, (of which this is not the proper place to give an account) I thought fit to make enquiry about the Vegetation and growth of Plants of confiderable Bulk in thofe fubmarine Regions, where if there grow any, they muft do it remote from the free contact of an ambient Air. And having not now the leifure to repeat what *Botanifts* (of whofe Books I am not now provided) deliver about leffer Plants growing under Water, I fhall now onely prefent you with what information I could procure from Navigators, about Trees and Fruit growing at the bottom of the Sea.

To

To what I have elfewhere had occafion to fay to their Opinion, that will not allow Coral to be really a Stony Plant, but a Livelefs Concrete, that is alwayes hard and brittle under Water; I fhall now add, that, inquiring lately of an Eminent and Inquifitive Perfon, that had fpent fome time upon the Coaft of *Africa*, where he had been prefent at the fifhing of Coral, and learning from his anfwer, that he had feen it not far from *Algiers*; I ask'd him, whether he had himfelf obferv'd the Coral to be foft, and not red, when 'twas newly brought from the bottom of the Sea. To which he replied, that he had found it foft and flexible; and that, as for the colour, it was for the moft part very pale, but with an eye of red, the Bark being worfe coloured than the fubftance it cover'd was; but when the Bark was taken off, and the other part expofed to the Air, the expected rednefs of the Coral difclos'd it felf.

When I demanded, whether he had obferved, that any inky fap afcended to nourifh the ftony Plant? and whether he had feen any thing like Berries upon it? He ingenuoufly confeffed to me, he had not been fo curious as purpofely to make inquiry into thofe Particulars; but that he remembred, That having broken fome of the large pieces of Coral, he took notice, that the more internal Subftance was much paler than the other, and very whitifh, and that at the extream parts of fome branches or fprigs he obferv'd little blackifh knobs, which he did not then know what to make of: And when I enquir'd, what depth the Sea was of in that place? he anfwer'd, that 'twas nine or ten fathom. But as to the Fruit of fome kinds of Coral, if I do not much mifremember, I was, not long fince, affured by a Scholar that navigated much in the Eaft, that they divers times meet with in thofe Seas a certain fort of Coral, but not white, which bears a fmall Fruit like a round Berry, of a pleafant colour, and efteem'd as rarities.

Difcourfing with a perfon that made *Diving* his Trade, whether he had not met with any Trees or Fruit in the depths of the Sea? He told me, that in a great Ship, whereinto he defcended, to recover thence fome fhipwrack'd Goods, he was furpriz'd to find in feveral places a certain fort of Fruit, that he knew not what to make of; for he found them of a flimy and foft confiftence, about the bignefs of Apples, but not fo round in fhape, and when he brought them up into the Air, as he did many of them, they foon began to fhrink up like old rotten Apples, but were much harder, and more fhrivel'd. And 'tis remarkable, that this happen'd in a cold Northern Sea.

One that made a confiderable ftay about *Manar*, a place I have often mention'd, anfwer'd me, that he learn'd from the *Divers*, that in fome places thereabouts there grows at the bottom pretty ftore of a certain fort of Trees, bearing Leaves almoft like thofe of Laurel, as alfo a certain Fruit; but of what virtue, or other ufe, he had not the Curiofity to enquire.

I was alfo inform'd by an Eye-witnefs, that near the famous Coaft of *Mofambique* in *Africa*, there grows at the Bottom of the Sea ftore of Trees, that bear a certain Fruit, which he defcribes

to be very like that, which in *America* they are
wont to call *Acayu*, the Leaves alfo refembling thofe
of that Tree.

But the welcomeft Information I could procure
about Sub-marine Plants, is that which concerns
the famous *Maldivian* Nut, or *Coco*, which is fo
highly efteem'd in the Eaft, that fome write, it is a
great Prefent from one King to another, and even
much extoll'd in *Europe* by experienc'd Phyfici-
ans : For the Origine of this dear Drug is almoft
as much controverted as the Alexiterial Virtues
are extoll'd. Having then once the good fortune
to meet with a man of Letters, that had refided in
thofe unfrequented Iflands, I found he had been
as inquifitive as I could reafonably expect about
thefe admir'd Productions of the Sea, and that
he had often learn'd from the *Divers*, that they
are real Nuts or Fruits born by a fort of Coco-
Trees that grow at the Bottom of the Sea, and
are thence either torn off by the agitation of
the Water, or gather'd by the *Divers.* Thefe
Fruits are fmaller than moft other forts of *Coco's,*
whofe maturity they do not feem to arrive at. He
thinks, the *Species* may have been very differing
from what it is, and may have come from Nuts
fallen into the Sea, together with the ruine of
fome little Iflands undermin'd by the Water, and
fo fubmerg'd ; of which, he told me, he faw at
leaft three or four inftances during his ftay there.
He told me, that whilft the Fruit was under Water, they obferv'd no diftinct fhell and kernell, but
the entire Nut was fo foft, that it may be eafily
enough cut with a knife, and was eaten like
their other Fruits ; but being kept about a Week

in

in the hot Air, it grows folid, and fo hard as to require good Steel Tools to work upon it. He added, that though even upon the place the fairer fort be of very great efteem, yet not of any fuch prodigious price as is given out. And he prefented me one about the bignefs of a large Egg, and a Fragment of another, which are both very hard ; but as for their Virtues, I can yet fay nothing upon Tryal, for want of having had fitting Opportunities.

Other Obfervations made at the Bottom of the Sea may hereafter follow.

F I N I S.

A
PARADOX
OF THE
NATURAL
AND
PRETERNATURAL STATE
OF
BODIES,
Efpecially of the AIR.

By the Honourable *ROBERT BOYLE*.

ADVERTISEMENT.

AN attentive Reader will easily be perswaded by a couple of passages in the following Papers that it is only a Fragment. But though the Author, for certain Reasons, has for divers years suppress'd the other Discourses that belong to the same Treatise; yet he was content to let this come abroad without them; not only because, relating chiefly to the Air, it may fitly be consorted with those Papers concerning some Qualities of the Air, which it accompanies; but because 'tis hop'd, it may prevent, or put an end to, several unnecessary Disputes about the Natural and Forc'd Constitution of the Air (warmly agitated among Learned men,) by shewing them to be founded, some upon precarious suppositions, and more upon Vulgar Mistakes. ⸺ ⸺

OF THE
NATURAL
AND
PRETERNATURAL STATE
OF
BODIES,

Eſpecially the AIR.

I Know, that not only in Living, but even in Inanimate, Bodies, of which alone I here diſcourſe, men have univerſally admitted the famous Diſtinction between the *Natural* and *Preternatural* or Violent ſtate of Bodies, and do daily, without the leaſt ſcruple, found upon it *Hypotheſes* and Ratiocinations, as if it were moſt certain, that (what they call *Nature*) had purpoſely framed Bodies in ſuch a determinate ſtate, and were alwayes watchful that they ſhould not by any external Violence be put out of it.

But

But notwithstanding so general a consent of men in this point, I confess, I cannot yet be satisfied about it in the sence wherein it is wont to be taken. 'Tis not that I believe, that there is *no sence*, in which, or in the account upon which, a Body may be said to be in its *natural* state ; but that I think the *common* Distinction of a *natural* and *violent* state of Bodies has not been clearly explained, and considerately setled, and both is not well grounded, and is oftentimes ill apply'd. For, when I consider, that whatever state a Body be put into, or kept in, it obtains or retains that state according to the Catholick Laws of Nature, I cannot think it fit to deny, that, in this sence, the Body propos'd is in a *natural* state ; but then, upon the same ground 'twill be hard to deny, but that those Bodies, which are said to be in a *violent* state, may also be in a *natural* one, since the violence, they are presumed to suffer from outward Agents, is likewise exercised no otherwise than according to the established Laws of Universal Nature. 'Tis true, that when men look upon a Body as in a *preternatural* state, they have an *idea* of it differing from that which they had whilst they believ'd it to be in a *natural* state : But perhaps this difference arises chiefly from hence, that they do not consider the condition of the Body, as it results from the Catholick Laws setled among things Corporeal, and relates to the Universe, but estimate it with reference to what they suppose is convenient or inconvenient for the particular Body it self. But however it seems to me, that mens determining a Body to be in a natural or preternatural state has much more in it, either of casual, or of arbitrary, or both, than they are aware of. For oftentimes we

think

action of that Agent, if the Body were under any violence, 'twas exercis'd by usual, but often imma‑nifest Agents, though perhaps their Compulsion were not *less,* but only less *heeded.* And sometimes also no more is to be understood by a Bodies being forc'd from its Natural state, than that it has lost that, which it had immediately, or a pretty while be‑fore some notable change. Which Conjectures I shall now endeavour to confirm, but with great bre‑vity.

I have already shewn, that Matter being devoid of sense and appetite, cannot be truly and proper‑ly said to *Affect* one state or condition more than another, and consequently has no true desire to continue in any one state, or to recover it when once lost; and Inanimate bodies are such, and in such a state, not as the material parts they consist of, elect‑ed or desired to make them, but as the natural A‑gents, that brought together and rang'd those parts, actually made them. As a piece of Wax is uncon‑cern'd, whether you give it the shape of a Sphere, or a Cone, or a Pillar, or a Boat; and whether, when it has that form, you change it into any other; the matter still retaining without willingness or unwil‑lingness, because without perception, that figure or state which the last action of the Agents (your fin‑gers or instruments) determined it to, and left it in.

But this will be best understood, as well as con‑firmed, by particular examples. I need not tell you, ter being heated by the Fire, as foon as that adventitious Heat vanifhes, returns to its native coldnefs; and fo when, by an excefs of Cold, it is congeal'd into Ice, it does upon a thaw lofe that preternatural hardnefs, and recover the fluidity that naturally belongs to it : And the fame may be likewife faid of Butter, which, being melted by external Heat into a Liquor, does upon the ceffation of that Heat grow a confiftent body again. But perhaps thefe inftances will rather countenance our *Paradox* than difprove it. For as to the coldnefs whereto Water heated by the Fire returns when 'tis remov'd thence, it may be faid, that the acquired Heat confifting but in the various and brisk agitation of the Corpufcles of the Water by an external agent, it need be no wonder, that when that Agent ceafes to operate, the effect of its operation fhould ceafe too, and the water be left in its former condition, *whether* we fuppofe it to have been heated by the actual pervafion of the Corpufcles of the Fire, which by degrees fly away into the Air ; *or* that the Heat proceeds from an agitation imparted by the Fire to the Aqueous Corpufcles, which muft by degrees lofe that new agitation, by communicating it little by little to the contiguous Air and Veffel ; fo that, if the former agitation of the particles of the Water, were, as is ufual, much more languid than that of our Organs of Feeling, in which faintnefs of motion the coldnefs of Water confifted, there will be no need of any pofitive internal

form,

form, or any care of Nature to account for the Waters growing cold again.   This will be confirm'd by
the consideration of what happens to Ice, which is
said to be Water brought into a preternatural state
by an excess of Cold: For, I doubt, 'twill not be easily demonstrated, that in reference to the nature of
things, and not to our arbitrary *ideas* of them, *Ice* is
Water preternaturally harden'd by Cold, and not
Water Ice preternaturally thaw'd by Heat. For if
you urge, that Ice left to it self will, when the Frigorifick agents are removed, return to Water; I
shall readily answer, that, not to mention the Snow
and Ice that lyes all the Summer long unthawed upon the tops of the Alps and other high mountains ,
I have learn'd, by inquiry purposely made, from a
Doctor of Physick, who for divers years practised
in *Muscovy*, that in some parts of *Siberia* (a large
Province belonging to the *Russian* Emperour) the
surface of the ground continues more Months of the
year frozen, by what is call'd the natural Temperature of the Climate, than thaw'd by the Heat of the
Sun; and that a little beneath the surface of the
ground, the Water, that chances to be lodged in the
cavities of the Soil, continues frozen all the year ;
so that, when in the heat of Summer the Fields are
covered with Corn, if then you dig a foot or two,
perhaps less, you shall easily find Ice and a frozen
Soil : So that a man born and bred in the inland
part of that Country, and inform'd only by his own
Observation, may probably look upon Water as
Ice violently melted by that Celestial Fire, the Sun,
whose heat is there so vehement in their short Summer, as to ripen their Harvest in less time than in our
Tempelate Climates will easily be credited.

On

On the other fide we in *England* look upon melted Butter, as brought into a violent ftate by the Operation of the Fire, and therefore think, that when being remov'd from the Fire it becomes a confiftent Body again, it has but recover'd its Native Conftitution. Whereas there are divers parts of the *Eaft Indies*, and, I doubt not, of other hot Countryes, whofe Inhabitants, if they fhould fee confiftent Butter (as fometimes by the care and induftry of the *Europeans* they may do) they would think it to be brought to a preternatural ftate, by fome artificial way of Refrigeration. For in thofe parts of the *Indies* I fpeak of, (though not in all others) the conftant temper of the Air being capable to entertain as much of agitation as fuffices for fluidity in the parts of what in our Climate would be Butter, 'twould be in vain to expect, that, by being left to it felf in the Air, it fhould become a confiftent Body. And I have learn'd by diligent inquiry of Sea-men and Travellers, both *Englifh* and others, that were Eye-witneffes of what they told me, that, in divers parts of thofe hot Regions, Butter, unlefs by the *Europeans* or their difciples purpofely made in the Cold, is all the year fluid, and fold, or difpens'd, not as confiftent Bodies, by weight, but as Liquors, by meafure. To ftrengthen this Obfervation, I fhall add, what was affirm'd to me by a Learned man, that practis'd Phyfick in the warmer parts of *America,* namely, that he met in fome places with feveral Druggs, which, though they there feem to be Balfoms, as Turpentine, *&c.* are with us, and retain'd that confiftence in thofe Climates, yet when they come into our colder Regions, harden into Gums, and continue fuch both

Win-

Winter and Summer. On the other fide, inquiring alfo of a Traveller, vers'd in Phyfical things,
about the Effects of great Heat in the in-land part
of *Africa*, where he had lately been ; he told me,
among other things, that Raifin of *Jalap*, which,
when he carried it out of *England*, was of a confiftence not only dry but brittle , did, when, and a
while before, he came to *Morocco*, melt into a fubftance like Turpentine ; fo that fome of it that he
had made up into Pills, would no more at all retain
that fhape , but remain as it were melted all the
while he ftayed in that City, and the neighbouring
Countrey, though when he came back to the bor-
ders of *Spain*, it return'd to its former confiftence.
Which I the lefs wonder'd at, becaufe, having had
the curiofity to confider fome parcels of Gum *Lacca*,
(of which Sealing Wax is made) newly brought
afhore from the *Eaft Indies*, though it be a hard and
folid Gum , yet I found by feveral inftances, that,
paffing through the *Torrid Zone*, divers pieces of it,
notwithftanding the fhelter afforded it by the great
Ship it came in, had been, by the Heat of the Climate, melted, and made to ftick together, though
afterwards they regain'd their former Confiftence,
though not altogether their former Colour. And
on this occafion I fhall add, that I learn'd by inquiry from a particular acquaintance of mine, who
brought me divers rarities out of *America*, that having at the place where 'twas made, among other
things, furnifhed himfelf with a quantity of the beft
Aloes, he obferved, that whilft he fail'd through
very hot Climates, it was fo foft, that, like liquid
Pitch, it would often have fallen out of the wide-
mouth'd Veffel he kept it in, if he had not from
time

time to time been careful to prevent it. But when he came within a hundred Leagues of the Coast of *England*, it grew hard , and so continued, though this were in a very warm season of the year, being about the Dog-dayes.

For further confirmation of what has been hitherto discoursed , be pleased to consider with me that most obvious Body , the *Air*, or the Atmosphere we live and breath in. For though several Opinions and Argumentations are founded upon what their Authors call the Natural and Preternatural or Violent state of the Air, yet he that considers, shall find it no easie thing to determine, what state of the Air ought to be reputed its truly Natural state, unless in the sence I formerly told you I employ that expression in. I will not insist on the Heat and Coldness of the Air ; for, that being manifestly very differing in the heart of Winter, and in the heat of Summer, and in differing Regions of the Air, as at the top and bottom of high mountains, at the same time, and constantly in differing Regions of the Earth, as in *Barbary* and *Greenland*, 'twill not be so easie to determine what state is natural to the Air. But that only which I shall now consider, is its state or tone in reference to *Rarity* and *Density*. For, since the Air is believed to be condensed by Cold and expanded by Heat, I demand, at what time of the year , and in what Countrey, the Air shall be reputed to be in its *Natural state?* For, if you name any one time, as the Winter, or the Summer, I will ask, why that must be the standard of the tone of the Air rather than another Season, or at least exclusively to all others? And the like difficulty may be made about the Climate

mate or the Place. And thefe cruples are the more
allowable to be propos'd, becaufe Learred men
have deliver'd, that in fomé Countryes the *Mercu-*
ry in the *Torricellian* Experiment, is kept higher than
in others, (as in *Sweden* than in *Italy*,) and our Ba-
rofcopes inform us, that oftentimes, in the fame
place and day, the Quick-filver in the fame Inftrument does confiderably vary its height ; which
fhews, that the Air or Atmofphere muft neceffarily
vary its weight, and therefore probably its degree of
Rarity or Denfity.

But I have yet to propofe a further Confiderati-
on in this Affair : For, what if it fhall appear, that
neither in Winter nor in Summer, in *Sweden* or in
Italy, or in whatever Country, Region, or Seafon
you pleafe, the Air we breath in is in any other
than a Preternatural ftate; nay, that even when we
have vehemently agitated and expanded it by an in-
tenfe heat of the Fire, it is not yet violently rarified, but yet violently conftipated, unlefs in our
fence before declared, you underftand with me the
Preternatural ftate of Rarefaction in the Air, in reference to the tone it had before the laft notable
change was produc'd in it. This will, I queftion
not, feem a furprizing, if not a wild, Paradox : But
yet to make it probable, I fhall only defire you to
reflect upon two or three of my *Phyfico-Mechanical*
Experiments; and there you will fee, *firft*, that
the Air being a Body abounding with fpringy particles, not devoid of Gravity , the inferiour muft
be comprefs'd by the weight of all the incumbent.
And *next*, that this Compreffion is fo great, that
though by the heat of the Fire neither others nor
we could bring a portion of included Air to be ex-
parded

panded to above fourscore times its former space; yet without heat, by barely taking off the pressure of the superiour Air, by the help of our Pneumatical Engine, the Air was rarified more than twice as much : And since those Experiments were published, I more than once rarified it to above five hundred times its usual Dimensions; so that, if according to what is generally agreed on and taught, a Body be then in a Preternatural state, when by an external force it is kept in a condition, from which it incessantly tends to get free.; and if it be then most near its Natural state; when it has the most prosperously endeavoured to free it self from external force, and comply with its never-ceasing tendency; if this be so, I say, then the Air we live in is constantly in a Preternatural state of Compression by External force. And when it is most of all rarified by the Fire, or by our Engine, its Springs having then far more conveniency than before to display themselves, which they continually tend to do, it answerably approaches to its Natural state, which is to be yet less compressed or not at all. And I have carefully try'd for many months together, that when the Air has been rarified much more than even a vehement heat will bring it to be, yet if it were fenc'd from the pressure of the external Air, it would not shrink to its former dimensions, as if it had been put into a violent state, from whence Nature would reduce it to them, but continued in that great and seemingly preternatural degree of extension, as long as I had occasion to observe it. One might here shew, that this odd constitution of the Air is so expedient, if not necessary for the Motion, Respiration, and other uses of Animals, and in particular

ticular of men , that the Providence and Goodnefs
of the Wife Author of the Univerfe is thereby
fignally declared; if it were not improper in fuch
a Paper as this to imploy *final* Caufes. Wherefore
to avoid the imputation of impertinence, I will conclude, by taking notice that from what has been
delivered we may learn two things confiderable
enough, if not in themfelves, yet to fome paffages
of the Treatife, whereof this Paper makes a part.
And firft, we may deduce from what has been faid
of the Air, that according to what is noted above,
That may fometimes generally be granted and believ'd to be the Natural ftate of a Body, not which
it really affects to be in , or (to fpeak more properly) has a tendency to attain, but that which it's
brought into and kept in by the action or refiftance
of neighbouring Bodies, or by fuch a concourfe of
Agents and Caufes as will not fuffer it to pafs into
another ftate. And the fecond thing we may hence
learn is, that whatever men fay of Natures never
miffing her aim, and that nothing violent is durable; yet, bating an inconfiderable Portion of Aerial
particles at the upper furface, for ought we know
the whole mafs of the Air we live in, and which
invirons the whole Terraqueous Globe , has been
from the worlds beginning; and will be to its end,
kept in a ftate of violent Compreffion.

F I N I S.

Hygroscope

Propofed to be further tryed,

Together with

A BRIEF ACCOUNT

OF THE

Utilities of HYGROSCOPES,

By the Honourable *ROBERT BOYLE*,
Fellow of the *Royal Society*.

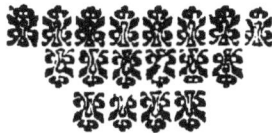

L O N D O N,
Printed by *E. F.* for *R. Davis*, Bookfeller
in *Oxford*, 1673.

Lyquicope

BRIEF ACCOUNT

A
STATICAL HYGROSCOPE

Propofed to be farther tryed,

In a Letter to *H. Oldenburgh* Efq;

Secretary to the

ROYAL SOCIETY.

SIR,

THough I writ to you from *Stanton* an account of thofe Hygrofcopes, whereof I now prefent you one; yet, fince I remember that it was in the year 1665 that I fent you that Paper, I fear you may by this time have forgotten much of what it contain'd, and thereby made it fit for me in this Letter, both to remind you of fome former paffages, and to add fome Obfervations that lately occurr'd to me; and this the rather, becaufe I do not prefent you with this trifle meerly to gratifie your Curiofity, but that you

L 3   and

and some of your ingenious friends may, by your remarks, help me to discover to what inconveniences our Instrument is liable, how far they may be avoided or lessen'd, or what the uses or advantages of it may be, notwithstanding its inevitable inconveniences or imperfections.

Having had occasion amongst other subjects relating to the Air, to consider its Moisture, and its Dryness, I easily discern'd that they had no small influence upon divers Bodies ; and among the rest, upon those of men, as the ambient Air we breath in, either passes from one of those Qualities to the other, or even from one degree to the other in the same quality.

Wherefore I began to cast about somewhat sollicitously for a way that might better than any I had yet tryed, or elsewhere met with, discover the changes of the Air as to moisture and dryness, and the degrees of either quality. For which purpose it seem'd to me, that, if a Statical Hygroscope could be had, it would be very convenient, in regard of its fitness, both to determine the degrees of the moisture or dryness of the Air, and to transmit the Observations made of them to others. Whereupon considering further, that among Bodies otherwise well qualified for such a purpose, that was likeliest to give the sensiblest informations of the changes of the Air, which, in respect of its bulk, had the most of its surface exposed thereunto ; I quickly pitch'd upon a fine Spunge, as that which is easily portable, not easie to be divided or dissipated, which, by its readiness to soak in Water, seem'd likely to imbibe the Aqueous particles that it may meet with dispers'd

fpers'd in the Air, and which, by its great poroufnefs throughout, has much more of *Superficies* in reference to its bulk, than any Body not otherwife lefs
fit for the intended ufe that came into my thoughts.

If you recall to mind, when and whence I firft
gave you notice that I employed our little inftrument, you will eafily believe, that the Inducements
I had to pitch upon it, were, that I fhould need
but fuch light and parable things, as I could eafily
both procure in the Country (where I then was)
and carry about with me in the frequent removes I
was obliged to make ; and therefore that I did
not reprefent this trifle as the beft Hygrofcope that
could be devifed, or even as the beft that perhaps
I my felf could have propounded, if I would have
fram'd an elaborate Engine with Wheels, Springs,
or equivalent Weights, Pullies, *Indices,* and other
contrivances, fome of which I divers years ago
made ufe of. For I little doubt, but that Mechanical
heads may frame Hygrofcopes much curioufer and
perfecter than that I now fend you, or any other
I have ufed or feen, if they may be accommodated
with fufficient room, and dextrous Artificers that
will work exactly according to directions; whereas
my defign being not fo much to make a *Machinal*
or Engine-like, as a *Statical, Hygrofcope,* and fuch
an one as may be fimple, cheap, contained and fet
up in a little room, eafie to be made and tranfport-
ed, I thought it might be of fome ufe, efpecially to
thofe that are not furnifhed with Curiofities and
Mechanical Accommodations, if among the feveral
forms of Hygrofcopes that I had in my mind, I
chofe one, that being *ftatical* and eafie, might be as

com-

commodious by its simplicity, as some others by their elaborateness; especially if we consider, that, as slight an instrument as it seems, it may be applyed to various uses, some of which are not slight, as will ere long be made probable.

If I should be here told by one, that grants the preferableness of Statical Hygroscopes in the general, that there are *divers Bodies*, other than that pitch'd upon by me, whose weight may vary when the Temperature of the Air is considerably altered as to dryness and moisture, and that perhaps among these, some one may be found that may imbibe the Aqueous particles of the Air better than our Spunge; I shall not resolutely deny it, and therefore shall leave you to make tryals with what other Bodies you shall think fit, contenting my self to have suggested in general the conveniency of making Hygroscopes, where the differing changes of the Air may be estimated by weight; but this I shall tell you in favour of our Spunge, that when I was considering, what Bodies were the fittest to be employed for the making of Statical Hygroscopes, I made tryal of more than one that seem'd not the least promising. I know, that Common or Sea-Salt will much relent in moist Air, and Salt of Tartar will do it much more; but then those Salts, especially the latter, will not so easily as they should, part with the Aqueous Corpuscles they have once imbibed, and are in other regards, (which 'twere not worth while to insist on,) less convenient than a Spunge. I made tryal also with Lute-strings, which were purposely chosen very slender, that they might have the greater surface in respect of

their

their bulk ; thefe I found at firft to do very well, as to the imbibing of the moifture of the Air, but afterwards they did not continue to anfwer my expectation. I caus'd likewife to be turn'd out of a light wood a Cup, which, that it might lefs burden a tender balance, had, inftead of a foot, a little button, to which a hair might be tied, to fufpend it by; and this Cup being purpofely turn'd very thin, that it might have much furface expofed to the Air, proved for a pretty while fo good a Hygrofcope, as invited me to make divers Obfervations with it, fome of the which I have ftill by me. It agreed alfo with feveral tryals, that I had made on other occafions, of the poroufnefs of fuch Bodies, that white Sheeps Leather, fuch as Chirurgeons us'd to fpread plaifters upon, would be very convenient for my purpofe. And indeed I found by many Obfervations, whofe fuccefs you may command a fight of, that if this Leather were a fubftance as little obnoxious to Corruption as a Spunge, it would, by its copious imbibitions and emiffions of the Aerial moifture, be a fitter matter than any other I had employ'd for a Hygrofcope.

But taking all things together, I found no Body fo convenient for my purpofe as a Spunge, which you will perhaps the more eafily believe, if I add, that to help me to make fome eftimate of the porofity of it, [We weigh'd out a dram of fine Spunge, and having fuffer'd it to foak up what Water it could, it was held in the Air, not only whilft the weight of the Water would eafily make it run out, but till it dropt fo very

slowly,

flowly , that a hundred was reckon'd after one drop before another fell ; then putting it into the balance it had been weighed in before , we found , that , as its dimenfions were increafed to the Eye, fo its weight was increas'd upon the fcale, amounting now to fomewhat above two Ounces and two Drams ; fo that one Dram of Spunge, though it feem'd not altogether fo fine as the portion we had chofen out for our Hygrofcopes, did imbibe and retain feventeen times its weight of Water.]

Now when one is refolved to employ a Spunge, there will not need to be much added about the turning it into a *Hygrofcope.* For, having weigh'd it when the Air is of a moderate Temperature, it requires but to be put into one of the fcales of a good balance fufpended on a Gibbet (as they call it) or fome other fix'd and ftable fupporter. For the Spunge being carefully counterpoifed at firft with a metalline weight (becaufe that alters not fenfibly with the changes of the Air) it will by its decrement or increafe of weight fhew, how much the neighbouring Air is grown dryer or moifter in the place where the inftrument is kept. The weight of the Spunge may be greater or lefs according to the bignefs and goodnefs of the balance, and the accuratenefs you defire in the difcoveries it is to make you. For my part, though I have for Curiofitie's fake with very tender fcales imployed for a good while but half a dram of Spunge, and I found it to anfwer my expectation well enough ; and though, when I us'd a

bulk

bulk divers times as great, in a stronger, but proportionably less accurate, balance, I found not the Experiment successless; yet after tryals with differing quantities of Spunge, I preferr'd, both to a greater and lesser weight, that of a dram, as not being heavy enough to overburden the finer sort of Goldsmiths scales, and yet great enough to discover changes considerably minute, since they would turn *discernably* with a sixteenth or twentieth part, and *manifestly* with half a quarter of a grain.

With such *Hygroscopes* as these (wherein the balance ought to be still kept suspended and charged) I made several tryals, as my removes and accommodations would permit, sometimes in the Spring, and sometimes in the Autumn, and sometimes also in the Summer and Winter. But nevertheless it would be very welcome to me, if you and some of your Friends would be pleased to make tryals your selves, and compare them with mine, and especially take notice, if you can, whether in any reasonable tract of Time there will be any loss (worth noting) of the substance of the Spunge it self; I having not hitherto discover'd any. In the mean time, to invite you to give your selves this trouble, after I have told you, that having once, among divers removes, had the opportunity to keep a dram of Spunge suspended during a whole Spring, and a great part of the preceding Winter and subsequent Summer, I did not think my pains lost, though divers of the observations they afforded me have unhappily been so, among many
ny

ny other memorials about Experiments of differing kinds; notwithſtanding which unſeaſonable loſs
I ſhall venture to ſuggeſt ſome things to you, that
occurr'd to me about the *Utilities* of the Inſtruments I am treating of.

A BRIEF

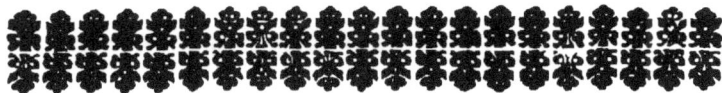

A
BRIEF ACCOUNT
OF THE
Utilities of HYGROSCOPES.

He ufe of a Hygrofcope is either general or particular : The former is almoſt coincident with the Qualifications to be wiſhed for and aim'd at in the Inſtrument it ſelf; The latter points out the particular applications that may be made of it when 'tis duely qualified. Of each of theſe I ſhall briefly ſubjoyn what readily occurrs to me.

The general uſe of a Hygroſcope is, *To eſtimate the changes of the Air, as to moiſture and dryneſs, by wayes of meaſuring them, eaſie to be known, provided, and communicated.*

I might here pretend, That as theſe are the principal things that have been deſir'd in Hygroſcopes, ſo 'tis obvious from the deſcription and account we

have

have given of our Inftrument, that thefe advantages belong to it in no very defpicable degree. And that to make fuch Hygrofcopes as will perform all thefe things in perfection, whatever it may feem to a Mental contriver, will, I fear, prove no eafie task to thofe that really attempt it. To thefe things I might add, that if fuch allowances be made, as what I have reprefented may invite you to grant, the Qualifications lately mentioned, as defirable in a Hygrofcope, may in a tolerable meafure be found in ours, when we fhall come to mention the particular ufes of it. And as for that of conveying to others the Obfervations made with it, you may pleafe to confider, that the things I employ to meafure the degrees of drynefs and moifture in the Air, being grains, parts of grains, and greater weights, the acceffions of moifture which the Spunge receives, or the loffes that it fuffers, can be eafily and at the fame time both found and determin'd. And as the weights imployed to determine thefe differences are eafily procurable; fo the Obfervations made with them, may (together with patterns, if it fhould be needful, of the weights themfelves,) with the fame facility be communicated by Letters even to remote parts. In which conveniency, whether, and how far, our Inftrument has the advantage of that made with an *Oaten beard,* and fome others that I have imployed, I leave you to confider.

I might farther alledge on the behalf of our Inftrument, that whereas, befides the Qualifications above mentioned, there is another, namely Durablenefs, which though not fo neceffary to conftitute a Hygrofcope, yet is neceffary, as will ere long appear, to fome of the confiderableft ufes of it:

And *whereas* such a Durableness is wished, as may not only keep the Instrument from having its substance rotted or corrupted by the Air, but may also preserve it in a capacity to continue pretty uniformly its Informations of the Air's moisture, even when that increases very much, or lasts very long; *Whereas*, I say, these things are much desir'd in a Hygroscope, our Spunge seems herein preferrable to the Oaten beard, Lute-strings, &c. For in those and the like Bodies the self-contracting or relaxing power (as 'tis suppos'd) or the disposition to imbibe and part with the moisture of the Air uniformly or after a due manner, is wont to be in no very long time alter'd or impaired; and particularly, when they have imbib'd much aerial moisture, they are very faintly affected by the supervening degrees of it, and so the operation is too disproportionate to what the like Cause would have produc'd, when the Instrument was well dispos'd; whereas in our Spunge neither the degree of springiness, nor any such like quality is consider'd, and it is capable of imbibing so much more of the Aqueous particles, than even moist Airs and Seasons are wont to supply it with, that there is little fear that it will be glutted, or have its pores choaked up with them, so that the decrements and accessions of weight will be more proportionate to the degree of moisture in the Air, and more reducible to known and determinate measures.

But though these and the like specious things may be represented in favour of our Statical Hygroscope; yet, to deal ingenuously with you, I much fear, that 'twill be very difficult to bring either Statical ones, or perhaps any other, to be so compleat as to satis-

fie

fie a nice and fevere Critick. And you would perhaps eafily aſſent to my Opinion, if it were not too tedious to entertain you with all the ſpeculative doubts and ſcruples, as well Mechanical as Phyſical, which my accuſtom'd diffidence has now and then ſuggeſted to me. But becauſe ſuch a ſceptical Diſcourſe would be too tedious, and alſo ſomewhat improper, to be propos'd by one that would recommend Hygroſcopes, I ſhall only now take notice of one great Imperfe45tion, which all that I have been acquainted with are liable to ; namely, that men have not yet found, nor perhaps ſo much as dream'd of ſeeking, a Standard of the Dryneſs and Moiſture of the Air, by relation to which, Hygrometers may at firſt be adjuſted, and ſo be compar'd with one another, as we ſee many of thoſe ſeal'd Thermoſcopes, that have been made and juſtn'd by Mr. *Shotgrave* the dextrous Operator of the *Royal Society.* I deny not, that, by virtue of a ſtandard to eſtimate moiſture by, I have endeavour'd to remedy this inconvenience ; but, as my hopes were but ſmall, ſo neither was my ſucceſs great, but I am not ſure, that happier Wits, or I my ſelf at ſome other and luckier time, may not more proſperouſly attempt it. In the mean while perchance you will not think it altogether nothing, if the Trifle I preſent you perform at leaſt ſome of the things deſir'd in a Hygrometer leſs imperfectly, than any you have yet met with. And that you may not be diſcourag'd by what I have lately acknowledg'd of the defects of ſuch Inſtruments, I think it now ſeaſonable to proceed to the mention of the particular *Uſes*, for which, notwithſtanding any inevita-

evita-

evitable defects, a Hygroscope, and even such a one as I now present you, may be made easily to serve.

USE I.

To know the differing Variations of Weather in the same Month, Day and Hour.

IT may be useful for divers purposes, to know both that the Air is wont to be less moist at one part of the Artificial Day (and so of the Night,) than at any other, & at what particular time of the Day or Night it most usually is so. And on this occasion I remember, that usually when the Weather was at a stand, it was observed, that the Spunge had manifestly gain'd in the Night, though it were kept in a Bed-chamber, and grew lighter again between the morning and noon. This Observation which was made towards the end of Winter would not hold, in case frosty nights or some other powerful Cause intervened. 'Twere not amiss also to observe, Whether there be not a Correspondence betwixt the Hygroscope and Baroscope ; and, if there be, in what kind of Weather or Constitution of Air it is most or least to be discerned. And this enquiry seems the more dubious, because the same changes of the Atmosphere may, upon differing accounts, have either the like, or quite contrary, operations upon these two Instruments. For in Summer when the Atmosphere is usually heavier, the Hygroscope is

M   usually

ufually lighter; fome ftrong Winds, as with us the North-weft, may make both the Atmofphere and Barofcope lighter, whereas Southerly Winds, efpecially if accompanied with rain, often make the Atmofphere lighter and the Spunge heavier. And on the other fide I obferve, that Eafterly Winds, efpecially when they begin to blow in Winter, though, by reafon of their drynefs, they are wont to make the Hygrofcope lighter, yet they are wont, at leaft here at the Weft-end of *London*, to make the Barofcope fhew the Air to be heavier. It were likewife fit to be obferved particularly by thofe that live on the Sea-coaft, Whether the daily ebbing or flowing of the Sea, do not fenfibly alter the weight of the Hygrofcope. It were very well worth while alfo to take notice, at what time of the day or night, *cæteris paribus*, the Air is the moft damp and moft dry, and not only in feveral parts of the fame day, but in feveral dayes of the fame month; efpecially on thofe days, wherein the full and new Moons happen. And this feems a more hopeful way of difcovering, whether the full Moon diffufes a moifture in the Air, than thofe Vulgar Traditions of the plumpnefs of Oyfters and Shell-fifh, and brains in the heads of fome Animals, and of Marrow in their bones, and divers other *Phænomena*, which, as I have fhewn in another paper, 'tis not eafie to be fure of. It may alfo be noted, whether Monthly Spring-tides, efpecially when they fall out near the middle of *March* or *September*, have any fenfible operation upon our Inftrument.

USE II.

To know how much one Year and Season is dryer or moister than another.

THis cannot be fo well perform'd by the Hygrofcope made of an Oaten beard, if they, that have made ufe of them more than I, do complain with reafon, that after fome months (for I cannot tell you precifely how many) they begin to dry up and fhrink ; fo that their fenfe of the varying degrees of the moifture of the Air is not fo quick as before, and the informations they give of the degrees of it, efpecially towards the outmoft bounds of their power to fhew the Air's alterations, recede more and more from Uniformity. But the lafting-nefs and other convenient qualifications of our Spunge making its capacity of doing fervice more durable, may the better help us to compare the greateft moifture and drynefs, both of the fame feafon, and of the feafons of one Year with the correfpondent ones of another. And if the weight of the Spunge at a convenient time, when the temperature of the Air is neither confiderably moift, nor confiderably dry, be taken for a Standard, a perfon that fhould think it worth his pains, may, by computing how many dayes at fuch an hour, and how much at that hour, it was heavier or lighter than the ftandard, and alfo by comparing the refult of fuch an account in one year with the refult of the like account in another year, be affifted to make a more particular and near eftimate of the differing temperature of

M 2

the Air , as to moisture and dryness, in one year than in another, and in any correspondent season or Month, assigned in each of the two years proposed. And how much the Collation or Continuance of such Observations , both in the same place and also in differing Countryes and Climates, may be of use to Physicians in reference to those Diseases, where the moisture and dryness of the Air has much interest ; and the Husbandman to fore-see what seasons will prove friendly or unkind to such and such Soils and Vegetables ; it must be the work of time to teach us, though in the mean while we have no reason to despair, that the Uses to be made of such Observations may prove considerable. And the rather, because if by help of the result of many Observations men be inabled to foresee (though at no great distance off) the temperature of a year, or even of a season, it may advantage not only Physicians and Plow-men, but other Professions of men, who receive much profit or prejudice by the dryness or excessive moisture of the seasons. And not to mention those who cultivate Hops, Saffron, and other Plants that are tender and bear a great price ; such a foresight, as we are speaking of, may be of great use to Shepherds, who, in divers parts of *England*, are oftentimes much damnified, if not quite undone, by the rot of Sheep, which usually happens through excess of moisture in certain months of the year. And in order to the providing of foundations whereupon to build Predictions, it may not be amiss to register the number, bigness, and duration of the considerabler spots, that may at this or that time of the year happen to appear or be dissipated on or near the Sun, or to take notice

of

of any extraordinary abfence of them , and to obferve whether their apparition or diffipation produce
any changes in the Hygrofcope : Which Curiofity
I fhould not venture to propofe, but that (as I elfewhere note) I find, that eminent Aftronomers have
cafually obferved great dryneffes to attend the extraordinary abfence or fewnefs of the Solar Spots.
And thofe perfons that are Aftrologically given,
may, if they pleafe , extend their Curiofity in the
ufe of this Inftrument to obferve, whether Eclipfes
of the Sun and Moon , and the great Conjunctions
of the Superiour Planets, have any notable operation
upon it.

U S E III.

*To difcover & compare the changes of the Temperature
of the Air made by Winds, ftrong or weak ; frofty,
fnowy, and other Weather.*

THis may conveniently enough be done as to
winds, either by our whole Inftruments or
(perhaps better and more fafely) by the Spunge
alone, which may be taken off and hung by a ftring,
for as long time as is thought fit, in the wind , and
then reftor'd to its former place.   For I found by
removing it into the wind, that it foon receiv'd a
very confiderable alteration in point of weight , as
alfo it did when remov'd out of a room into a garden where the Sun fhin'd ; for though the feafon
were not warm, it being then the Moneth of *January* ; yet in three quarters of an hour the fpunge loft
the 24th part of its weight.   We may alfo in fome

cafes

Cafes ufefully fubftitute to a Spunge a fomewhat broad piece of good Sheeps-leather difplay'd to the wind. For this having, by reafon of its thinnefs (or very fmall depth,) in proportion to its breadth, a very large *Superficies* immediatly expos'd to the wind, we found it to be notably alter'd thereby , in fo much that half an ounce of well prepar'd Sheeps-leather, (that we had long imployed as an Hygrofcope) being kept an hour in a place, where the Sun-beams might not beat upon it, did, in a ftrong wind, vary in that fhort time an eighteenth part of its original weight. But though I think it very poffible to make fuch obfervations of the Temperature of particular winds, as will frequently enough prove fo true as to be ufefull, at leaft to thofe that live in the places where they are made ; yet I am of opinion, that, to be able to fettle Rules any thing general, to determine with any certainty the Qualities of winds according to the corners whence they blow , as from the Eaft or Weft, North-eaft, South-weft, &c. there will be a great deal of warinefs requir'd ; and he that has not fome competent skill in Phyficks and Cofmography , will eafily be fubject to miftakes in forming his Rules. To countenance which advertifement, I fhall now make ufe but of thefe two Confiderations, whereof the firft is ; That winds that blow from the fame Quarter are not in fome Countryes of the fame Quality that they are in moft others, the wind participating much of the nature of the Region over which it blowes in its paffage to us. At the famous Port of *Archangel* they obferve, that whereas a Northerly wind almoft every where elfe without the Tropicks produces froft in Winter, there it is wont to be attended with a thaw, fo as to make the Eeves to drop. Of which the reafon feems to be, that this wind comes over the Sea which lyes North from that place; and on the contrary, a Southerly wind blowing over a thoufand or twelve hundred miles of frozen land does rather increafe the froft than bring a thaw. This was by the Inhabitants averr'd to the Ruffian Emperors Phyfician, who was more than once at *Archangel*, and from whom I had the Account. The Northern windes that are elfewhere wont to be drying, are faid in *Egypt* to be moift. I remember Mr *Sands*, in his exellent Travells, giving an account of what he obferv'd about the largeft of the fam'd Egyptian Pyramids, has this confiderable Paffage; *Yet this hath been too great a morfel for time to devour, having ftood, as may be probably conjectur'd, about three thoufand and two hundred years, and now rather old than ruinous: yet the Northfide moft worn by reafon of the humidity of the Northern Wind, which here is the moifteft.* Sands in *Purchas's Pilgrimage.*

And 'tis yet more confiderable to our purpofe what I find related by Monfieur *de Serres* in his ufefull book of Husbandry, fince by that it appears, that even in not very diftant Provinces of the fame Kingdome the winds that blow from the fame Quarter may have very differing Qualities and effects. For, fpeaking of the Changes of the Air in reference to Husbandry in feveral parts of *France*, he informes us, that 'tis obferv'd, that in the Quarters about *Tholoze* the South-wind dryes the ground, and the North gives rains. Whereas on the contrary from *Narbonne* to *Lyons*, all over *Provence* and *Dauphiné*, this laft nam'd wind caufes drynefs, and the other brings

Lib. 6. Cap. 8. Sect. 3.

Theat. d' Agricult. Lib. 1. chap. 7.

moift-

moisture.   And this may suffice for my first Consideration. My Second is this, That the vehemence
or the.faintness of the windes, though blowing over
the same country, may much diversify its operation
on the Hygroscope, and the same wind, which, when
it blows but faintly, or even moderately, is wont to
appear moist by the Hygroscope, may, when vehement or impetuous, make the Instrument grow
lighter, discussing and driving away more vapors
by the agitation of parts it makes in the Spunge,
than is countervail'd by those aqueous Vapors that
are brought along with it. But on such things as
these I have not leisure to infist, and therefore I shall
proceed to take notice in very few words of some
other operations of differing weathers on our Instrument, and tell you, that Frosty weather often
made the Hygroscope grow lighter even at night:
Snowy weather which lasted not long, added something to the weight of the Spunge. And it has been
observed that mists and foggy weather us'd to add
weight to it, even notwithstanding Frost.

To which may be added an Observation made by
my Amanuensis, who having a convenienter chamber than mine, (wherein a fire was daily made,)
was diligent and curious to set down the changes of
the Hygroscope that was left in his lodging ; for
this observation makes it probable , that a transient
cloud in fair weather may be (for I say not, that it
always is) manifestly observable by our Instrument.
For by his Diary it appears, that the 9th. of *September* being for the most part a very fair Sunshiny day, though about ten a clock in the morning
the Sun shone brightly, the Spunge began to preponderate, which unexpected *Phænomenon* made
                                    him

him look out at the window, where he difcover'd a
cloud that darken'd the Sun, but after a while that
being paft the balance return'd to an *Æquilibrium.*
On this occafion I fhall intimate, that I have more
than once or twice obferv'd, efpecially in Summer,
that when the Air grew heavier, the Hygrofcope
either continued at a ftand, or perhaps, alfo grew
lighter; as if, when fuch cafes happen, the *Effluvia*
that get into the Air, either from the Terreftrial or
fome other mundane globe, were not fit like vapors
to enter and lodge in the pores of the Spunge, and
fo were Corpufcles of another nature, with which
when we find by the Barofcope that the Air is
plentifully ftockt, it may be worth while to obferve,
Whether any, and if any, what kind of Meteor, as
Wind,or Rain it felf, or Hail, or in the Winter Snow
or froft, will commonly be fignified and produc'd.

U S E   IV.

To compare the Temperature of differing Houfes and
differing Rooms in the fame Houfe.

AS this is of great ufe both in refpect of mens
Health, efpecially if they be of a tender or
fickly conftitution, and in refpect of conveniency
for the keeping flefh, fweet-meats and feveral forts
of wares and goods, and even houfhold-ftuff, that are
fubject to be indammaged by moift air; fo it is
readily and manifeftly derivable from our Inftrument.   For, by removing it into feveral Houfes or
into feveral parts of the fame houfe, and letting it
　　　　　　　　　　　　　　　　　　　　　　ftand

ſtand in each a competent time to be affected with
the temperature of the Air of that particular place,
we have divers times obſerved a notable difference,
as you may gueſs by the two or three Notes I met
with among ſome old papers.

oᵇ. 13.   [Three or four days agoe a piece of fine
Spunge being taken out of a Cabbinet and clipt till it
came to weigh juſt half a drachm in a nice pair of
ſcales and a warm room, was afterwards remov'd
into a neighbouring room deſtitute of a chimney,
(and yet within 3 or 4 yards of a chimney ſel-
dom without fire :) This ſtatical Hygroſcope, conſiſting of the ſcales and the frame they hung on,
was yeſterday night remov'd into the former room,
and the Spunge was found to have gained 3 grains
and an eighth or better, and conſequently more than
a tenth part in reference to its firſt weight ; but being ſuffer'd to ſtand in this warm room, in leſs than
12 hours it loſt a grain and about ⅛ of its former
weight, though the time it ſtood in this room were
for the moſt part night and rainy weather.]

[We took a piece of very fine Spunge, which
formerly had weigh'd juſt a drachm, but having been
many months kept in a very warm room where fires
were kept every day, it was grown much lighter ; for,
removing it into an upper chamber in a neighbouring
houſe and weighing it in tender Scales, in the Evening 'twas found to want of a Drachm 4 grains and
¾ of a grain ; and though there was a fire in the room
and the Scales ſtood not far from it, yet, in a ſhort
time, (the day being foggy and rainy,) the Spunge
viſibly depreſs'd its Scale ⅜, and the next morning
was found to want but one grain and a half of a
Drachm, ſo that it had gain'd about three grains and

a quarter, and the following evening, being the fecond of *January*, it weigh'd one drachm a grain and almoft half a grain. So that in about one natural day the Spunge had acquir'd fix grains from the moifture of the Air, that is, a tenth part of its firft weight (I mean a drachm) and a greater proportion in reference to the weight it had the day before. The third of *January*, the weather being yet moift, the weight exceeded two grains, but about 3 or 4 of the Clock in the afternoon it began to lofe of that great weight, which diminifhed more by the next morning, the weather having chang'd that night and become fomewhat frofty.]

In another paper I alfo find this Note. [The drachm of a fpunge, that had for divers weeks been kept in a dry room, was (*January* the tenth) carried out into a room where fire is not wont to be kept, the weather being extraordinarily foggy: This morning being brought into the former room, though now the weather be clear (yet not frofty) it appears to have gain'd in weight about eleven grains; yet it foon loft 2 grains by ftanding in this room all the while in the balance.]

USE V.

To observe in a Chamber the effects of the presence or
absence of a fire in a Chimney or Stove.

THis is easily done, and the more easily if the
room be small. For in such chambers I have
often observed a moderate fire to alter the weight of
the instrument, placed at a distance from it, after it
had been well kindled but a very little while ; but
in wet weather, if the fire were not seasonably re-
newed with fresh fuel, the decay of it would, in no
long time, begin to be discernable by the Instrument.

USE VI.

To keep a Chamber at the same degree, or at an assign'd
degree, of Dryness.

SUppofing the alteration of weight in our spunge
to depend only upon the degree of the moisture
of the Air, the last named use will be but an obvious
Corollary of the former. For, if a convenient part
of the Room be chosen for the Hygroscope, and it
be kept constantly there, 'tis easy, by casting ones eye
on it from time to time, to perceive when 'twill be
requisite to increase or moderate the fire, so as to
keep the spunge at that weight it was of, when the
temperature of the Air of the chamber as to dryness and moisture was such as was desired. I will
not

not trouble you with some scruples, which I confess the consideration of this use of our Instrument suggested to me, because I have not now the leisure to discuss them. I had thoughts to try, whether and how far a good Quantity of salt of Tartar or even dryed Sea-salt, being kept in a closet or some closer room, might by imbibing lessen the moisture of the Air in it, but I did not perfect any observation of this kind. But I will add to what I have already referred to this sixth Head, that I have sometimes noted with pleasure, how manifest and great a change in the weight of our Spunge would be made, when the room was washed and a good while after, notwithstanding that a good fire was kept in it to hasten the drying of it.

Besides the hitherto mention'd uses of our Hygroscope, I know not whether there may not be divers others, and whether we may not, by a little altering and helping it, make it capable of shewing us some difference betwixt steams of differing natures, as those of Water, spirit of Wine, Chymical Oils, and perhaps new kinds of substances (such as we have not yet taken notice of) in the Air, in which, I confess, I suspect there may sometimes be dispersed store of Corpuscles, that I do not yet well know what to think of. For I have more than once observed (not without some wonder) the Hygroscope not to be affected with the alteration of weather, answerably to what the manifest constitutions or variations of it seem plainly to require : Whether unobserv'd Corpuscles perform'd this by making the other steams in point of figure, or size, incongruous to the minute pores of the spunge, and so unfit to enter them ; or by dissipating or otherwise pro-

procuring the avolation of more of the watery particles than they could countervail, I now examine not. And I am not sure, but by affociating this inftrument with the Thermofcope, Barofcope and

See ufes the
and the

fome others that may be propofed, it might be fo improved, as to help us to forefee divers confiderable things, that either are themfelves changes of the Air, or are wont to be confequences of them : As fickly and healthfull conftitutions of the Air both as to Man and Cattle, and healthful, barren or plentifull feafons in particular places or Countrys ; and perhaps alfo ftrong Hurricanes, Earthquakes, Inundations, and their ill effects, efpecially thofe accidents that depend much upon the furcharge of the Air, with other Exhalations and moift Vapors, which operate before fenfibly upon our Inftrument, and therefore may be difcernable by it a good while before they arrive at that height that makes them formidable Meteors. And if it were but the foretelling approaching rain, this very thing may on divers occafions prove very ferviceable, and recommend our inftrument, which often receives much earlier impreffions from the fteams that fwim up and down in the Air, than our fenfes do, fo that I have been able to forefee a fhowr of rain, efpecially in dry weather, a not inconfiderable while before it fell.

And here I fhould difmifs our fubject, which I have already dwelt on longer than I defign'd, but that remembring a caution I gave you when I was fpeaking of winds, I think it but fit to add two or three lines, to keep you from being

See the III ufe.

by that Advertifement difcouraged from endeavouring to make in the
ge-

general fuch Hygrofcopical obfervations, as may be reduc'd to *Hypothefes.* For, as I elfewere difcours'd concerning *Barometrical* Theories, if I may, fo call them; fo I fhall here reprefent concerning *Hygrofcopical* ones, that if a Theory or *Hypothefis* that is it felf rational, be found agreeable to what happens the moft ufually in obfervation; it ought not lightly to be rejected or fo much as laid afide, though fometimes we find particular Inftances, that feem to call it in queftion. For 'tis very poffible, that the Theory or *Hypothefis* may be as good as a wife man would require about fo mutable a fubject as the weather. And the Caufe affign'd by the *Hypothefis* may really act fuitably to what that requires, though a contrary effect infue by reafon of that Caufes being accidentally mafter'd and overrul'd by fome more powerfull Caufe or Agent that happens for that time to invade the Air. As we know that Tides do for the main correfpond with the motions of the Moon, (whofe *phafes* are therefore argued from them,) and do generally ebb and flow at fuch times and in fuch meafures as the Theory, that has been grounded on that correfpondency, requires; but yet Seamen find, that in this or that particular harbor or mouth of a River, fierce contrary winds, great Land-floods and other cafually intervening Caufes, do fometimes both very much difturb the regular courfe of the Tides, and increafe or leffen them.

F I N I S.

A NEVV
EXPERIMENT
And other INSTANCES
OF THE
EFFICACY
OF THE
AIR'S MOISTURE.

Subjoyn'd by way of

APPENDIX

To His

STATICAL HYGROSCOPE.

By the Honourable *ROBERT BOYLE.*

LONDON,
Printed by *E. F.* for *R. Davis,* Bookſeller
in *Oxford,* 1673.

ADVERTISEMENT.

THe *Author had thoughts of illuſtrating the fore-going Paper with a Collection of Hygroſcopical Obſervations, but though he ſeveral times begun Diaries of Occurrences of this Nature, as his Removes and other Avocations would permit ; yet beſides that theſe Impediments made him more than once break off his work, after he had continued it for a Month or two or longer, ſuch unwelcome Accidents happen'd ſince the foregoing Tract was ſent away, that he could not ſeaſonably recover any competent number of Obſervations, and fears he ſhall never recover ſome of them, which he doubts not to have been, with many better upon various ſubjects, ſtoln away from him. Upon which occaſion he thought fit to try, whether the following Paper might not be look'd upon as ſome amends for the miſſing of thoſe Obſervations in whoſe room it is ſubſtituted.*

A NEVV
EXPERIMENT
And other
INSTANCES
OF THE
EFFICACY of the AIR'S Moisture.

S Ince it may probably serve to recommend Hygroscopes to you, if that Quality of the Air, which these Instruments are usefull to give us an account of, be made appear to be more powerful, and have considerabler effects, than is commonly believed ; it will not be from my purpose to present you here some Instances that have led me to think, that the Effects of the *Moisture* of the *Air* may be considerable not only upon mens Healths, but upon subjects far less tender, and less curiously contriv'd, than Humane bodies. But I hope, you will easily believe, that by the Moisture of the Air I mean not a meer and

ab-

abstracted quality, but moist Air it self, or rather those humid Corpuscles, (chiefly of an Aqueous nature,) that abound, and rove to and fro, in our common Air.

That the Moisture of the Air may have no small influence, and usually a bad one, upon mens healths, is that, which, though Experience did not so often teach us, I should venture to argue from what I have observ'd of the operation of moist Air upon the dry and firmly context parts of Animals, and even in those cases, where, for want of time or other Impediments, this Moisture cannot produce any sensible degree of putrefaction.

That the skins of Animals may be easily invaded by the moist particles of the Air, is the more probable, because of the numerousness of their Pores, which may be concluded from their hairiness, or their sweat, or both. And I formerly observed to you, that I found *Sheeps-Leather* to imbibe the moisture of the Air, and increase in weight upon it, as plentifully as almost any Body I expos'd to it.

But to shew you, that much closer Membranes, and which Nature made to be impervious to such a Liquor as Urine it self, may be affected by the Vapours of the Air, I shall add, that having purposely taken pieces of Bladders fine and well blown, and, as far as appear'd, of a very close contexture, and counterpois'd them in a good balance, I found, according to expectation, that they would considerably increase their weight in moist, and lose it again in dry, weather; so that I might have employed the most membranous part of a bladder (for I thought not fit to make use of the neck or the adjoyning part) to make a *Statical Hygroscope.*

And,

And, as for other membranes and fibres, I ſhall have by and by occaſion to take notice, that even when they are ſtrongly and artificially wreathed together into gut-ſtrings, they may imbibe enough of the moiſture of the Air to be broken by it. And, I remember, I formerly told you, that I had obſerv'd *Lute-ſtrings* to grow heavier in moiſt Air.

And whereas *Bones* are by all confeſs'd to be the firmeſt and ſolideſt parts of Animals, and as it were the pillars by which the fabrick is ſuſtain'd ; yet it ſeems, that even they may be pierc'd into, and ſenſibly affected, by the moiſture of the Air. For I remember, that having cauſed the *Skeleton* of a humane body to be ſo made by a famous and very skilful Artiſt, that, by the help only of ſlender wires artificially order'd, the motions which the Muſcles make of the bones of a living body might be well imitated in the *Skeleton*, I obſerved, that though in dry and fair weather the flexures of the Limbs might be readily made, yet in very moiſt weather the joynts were not eaſily bent, as if the parts were grown ſtiff and rigid ; which ſeem'd to proceed hence, that moiſt particles of the Air, having plentifully inſinuated themſelves at the Pores into the Bones, had every way diſtended them, and thereby made the parts bear hard againſt one another, (which they did not at all before) at the Junctures or Articulations.

But it will be the more readily believed, that the Moiſture of the Air may operate conſiderably upon the tender and curiouſly contriv'd Bodies of Men and other Animals, if, proceeding to the Obſervations I chiefly deſign, I make it appear, that the moiſtning Particles, that rove up and down in the

Air,

Air, are able to exercise a notable (and, if I may so call it, a Mechanical) force even upon Inanimate and Inorganical bodies : which may well suggest a suspicion, that *Hygroscopes* being the proper Instruments to discover a Quality in the Air, whose efficacy reaches farther than is commonly taken notice of, they may in time be found useful to divers other purposes, besides those that relate to the health of men.

That *Wood*, especially when it has been season'd, is a Solid of a strong and firm contexture, if it were not obvious by the daily use made of it in building Ships, Houses, &c. might be easily concluded from the weight or force requir'd to alter its contexture by making any considerable, or perhaps sensible, Compression of it. And yet, that *Wood* may suffer a kind of divulsion of a multitude of its parts, and be manifestly distended by aqueous Corpuscles getting into its Pores, I remember, I proved by this Experiment. I got a piece of sound and season'd Wood of about an inch (or an inch and half) in Diameter, to be by a skilful Artist made Cylindrical, and also a ring of some solid matter, as Brass or Ivory, to be exactly turn'd to fit this Cylinder, so that it might without much ease, or much difficulty, be put on and taken off again : Then we put the turn'd piece of Wood into fair Water, and left it to soak there for many hours ; at the end of which it was visibly swell'd, and though I cannot now tell you, (for want of a Paper concerning that Experiment,) *how much* it was increas'd in Diameter, yet I well remember the increment was considerable, and that the ring, that was adjusted to it before, was manifestly too

little

little to be put again upon it, or with its Orifice to cover the whole *basis* of the diftended Cylinder, which afterwards being dryed in the Air fhrunk into a capacity of entring the ring again. And in this Experiment I took notice, that the great Intumefcence of the Wood was not produc'd all at once, or foon after it was put into the Water, but it fwell'd by degrees, and lay foaking there many hours before it arriv'd at its utmoft diftenfion, the aqueous Corpufcles requiring, it feems, fo much time to infinuate themfelves fufficiently into the Wood; which argues, that the internal parts were likewife affected, though, when even they came to fwell, they had a good thicknefs of Wood about them to hinder their Dilatation.

I expect you fhould now tell me, that this diftenfion of fo firm a Body was made by Water it felf, and not by the humid Vapours of the Air. On which occafion I might reprefent to you, that by the fweating (as men commonly call the adhefion of waterifh drops to the furface) of polifhed marble and fome other cold and fmooth Bodies, that fometimes happens even in the Heat of Summer, if they be cold, and the ambient Air moift enough; it appears, that both in hot weather the Air may be plentifully ftock'd with aqueous Vapours, and that thefe Vapours need to do no more than convene together to conftitute vifible and tangible Water. And on this occafion, if I were fure I had not told you of it already, I fhould fubjoyn an Experiment which would detect the Vulgar error of thofe that think the adhering drops, lately mention'd, to come from fome internal moifture deriv'd by its preffion or percolation from the marble or the

other

Body they are faften'd to ; and at the fame time I fhall fhew (what is not wont to be imagin'd) that in the Heat of Summer the Air is furnifhed with invifible and yet aqueous fteams. The Experiment I long fince try'd in Winter with Snow and Salt, included in a glafs Veffel, and then put to diffolve in a balance. But becaufe neither Ice nor Snow is at all eafie to be come by among us in *England* in Summer, and becaufe, at that feafon, the Air in fair weather is prefum'd to be dry as well as hot, I chofe, within fome dayes of *Midfummer*, and in clear Sun-fhiny weather, to make the following tryal.

We took a pint glafs-bottle, and having put into it a convenient quantity of Water (for room muft be left for the Salt) we plac'd them and four ounces of beaten *Sal Armoniack* in one fcale of a good balance, and a counterpoife in the other, and then, putting the Salt into the Water, I obferv'd, that though for a while the *Æquilibrium* remain'd, yet when the frigorifick mixture had fufficiently cool'd the outfide of the Bottle, the roving Vapours of the Air, that chanc'd to pafs along the furface of the Veffel, were, by the contact of that cold Body, arrefted, and turn'd into a kind of a dew , which from time to time made the fcale, that held the glafs, preponderate more and more, and at length the drops growing greater and greater, ran down in fmall rivulets the fides of the Glafs, and in lefs than an hour, (by my eftimate,) the condens'd fteams amounted to near a dram, which weight was afterwards much increas'd within about two hours more ; Whereby it fufficiently appears, both that this dew came from without, (fince if it had been a tranfudation,

tion, it would not have added weight to the fcale that received it,) and that there is even in clear Summer weather a vaft number of moift particles difpers'd through the Air, fince, in about an hours time, fuch a multitude of them as the Liquor pro-duc'd may be fuppos'd to confift of, and may by Heat be actually refolved into, could in courfe come to touch fo fmall a furface, as that of that part of fo fmall a bottle which contain'd the frigorifick mixture. For the reft of the Veffels furface was not cold e-nough to condenfe the Vapours into Liquor. But to return to what we were faying of Wood fwell'd by water; becaufe, notwithftanding thefe Confide-rations, I am willing to allow, that the Experiment of the Cylinder does not fully come home to our purpofe, and that I produc'd it not fo much to *prove directly* the force of moift Air, as to *countenance* what I am about to fay, by fhewing what a fufficient number of aqueous Corpufcles may do in the folid wood they penetrate, I fhall now add fome inftan-ces of the force thefe particles may exercife upon Solids, when they invade them but in the form of Vapours.

That in this form the multitude, figures, and motions of thefe infinuating particles may inable them to difplay no fmall force in their operations on fome Bodies, we have one Inftance that often hap-pens, though but feldome reflected on, in the break-ing of the ftrings of Mufical Inftruments, firft brought to a good Tenfion, upon the fupervening of rainy weather. For the caufe feems to be, that the Vapours that then wander through the Air, infinu-ating themfelves into thefe ftrings, (which the Mu-fician often forgets to let down or relax after having

ving skrew'd them up,) diftend and fwell them,
and thereby endeavour to fhorten them, and that
fo forcibly, that they not feldome break with a fmart
noife and great violence; which, becaufe it happens without any vifible efficient, men commonly
think and fay, that fuch ftrings break of themfelves.
But, to take no further notice of this popular fur-
mize, if we confider how much weight fome of thofe
bigger ftrings, efpecially of Bafe Viols, that have
been obferved to break in rainy weather, will require to ftretch any of them to a rupture, you will
eafily be induc'd to think that this operation of the
moift Air exacts, and therefore argues, more than a
languid force.

But here probably you will tell me, that the Inftances you expected were concerning Wood, which
is a far folider Body than gut-ftrings. To this I
fay, that the newly recited Inftance belongs directly to the title of this Paper, and, being above referr'd to, ought not to be pretermitted. And, as to
your expecting Inftances concerning Wood, I
might content my felf to refer you to what is obferv'd about the uneafie opening and fhutting fome
doors well adjufted to the door-cafe in very rainy
weather. But though this Obfervation favours my
defign, yet I had rather give you Inftances in wood
purpofely and carefully feafon'd. And therefore I
fhall now inform you of thefe two things; *one*
that I found by tryal (as I have elfewhere noted)
that Wood counterpoifed in a good balance would
grow fenfibly heavier in wet weather, and lighter a-
gain in dry; and the *other*, that, to fatisfie my felf
yet further, I confulted an ancient Mufician, to whom
I had once been a Difciple, and a famous Organ-
maker,

maker, to know whether they had not obferv'd
that the wood it felf, *&c.* of Mufical Inftruments
would receive fuch alterations from the moifture of
the Air, as might be difcern'd by the Ear? Upon
which inquiries, the Mafter of Mufick anfwer'd me,
That though Metalline ftrings will not change with
the weather like Gut-ftrings; yet Virginals (for
inftance) though furnifhed with wire-ftrings, will
for the moft part of them, (for fome he has obferved to be fo well feafon'd that they are not alter'd
by the weather,) be out of Tune in wet weather,
the ftrings generally then affording their notes
fharper than they fhould or are wont to do. And the
Organ-maker confefs'd to me, that, upon great
changes of weather, divers Organs would (after they
had been long ago tuned) grow out of tune, and that
not only the woodden pipes would be thereby
fwell'd, but the Metalline pipes untuned.

But if Bodies be of fuch a Conftitution as not only
to admit but affift the operation of the moift Air,
the penetrancy and efficacy of this may be found
much more confiderable than in the fore-going Inftances. For there are fome kinds of thofe Marchafites that yield Vitriol, which, whilft they lye under
ground, or are cover'd with the Sea-water, on
whofe fhores they are in fome places to be found,
retain a ftone-like hardnefs, and are often taken for
meer ftones; and yet fome credible perfons that are
converfant about Vitriol have cafually obferv'd, that
thefe, being expos'd to the Air, would in tra¢t of
time be fo penetrated by the moift particles of it,
though perhaps not meerly as moift, that (probably
by the help of the Vitriolate Corpufcles they met
with among the ftony matter) thefe hard and folid
Mar-

Marchafites are brought to fwell fo much as to burft.
That this will happen to fuch kind of ftones (though
they be of a clofe and heavy nature) by the help of
rain, Experience has perfwaded me, and that it may
alfo happen even to very hard and ftone-like Marchafites, (for many are not fuch,) when they are
meerly expos'd to the Air, I am apt to think upon
fome tryals of my own. For from fhining Marchafites, though but kept in my Chamber window, I have
had Vitriolate Efflorefcencies that feem'd to be produc'd by the action of the piercing moifture of the
Air upon the Mineral. And I remember, that very
hard and heavy lumps that were of a Marchafitical
fubftance, (though not at all gliftering,) which
feem'd to be ftony, were fo difpos'd to be wrought
on by the Air, that though they were kept partly in
my own chamber, and partly in other cover'd places, yet in no very long time they were fo penetrated by the moift Corpufcles of the Air, that they
were not only burft, but broken into many pieces;
infomuch that many of them did of themfelves fall
off from one another, and feveral of the divided portions would eafily be crumbled betwixt ones fingers.
And of fome of thefe I have obferved with pleafure,
that a Vitriolate fubftance was produc'd more copioufly in their innermoft parts than on or near their
outfide. So that, when I confider'd how great an external force would have been requifite to make fuch
a Comminution of Minerals fo folid and hard, 'twas
obvious for me to look upon the Air's moifture, as
capable, when it meets with fitly difpos'd Bodies,
to exercife a far greater force than is wont to be
conceived.

To thefe *Phænomena* I might add fome others to
the

the ſame purpoſe ; but becauſe the Marchaſites, and other Bodies requir'd to the producing of them, are not eaſie to be come by, and the ſucceſs often exacts a good length of time, I ſhall conclude this Paper by ſubjoyning a far ſhorter Experiment, that I devis'd not only to ſhew in general, that the moiſture of the Air may have a conſiderable Efficacy, but to aſſiſt a *Virtuoſo* to make ſome eſtimate in known meaſures of the Mechanical force of the Aerial moiſture. And though I now find to my trouble, that I want ſome of the Notes that concern the Circumſtances and the progreſs of the tryal, yet enough having eſcap'd to furniſh me with the following account of it, what I ſhall ſet down may, I hope, at leaſt put you in the way of repairing my misfortune.

Thinking it then probable, that Ropes themſelves would conſiderably imbibe and diſmiſs the moiſture of the Air, and that ſo as to ſhrink in rainy weather, though clogg'd with a weight faſten'd at the lower end, I was not diſcourag'd from attempting the following Tryal, by conſidering that the weight would ſtretch the Rope, and conſequently hinder the preſum'd effect of the Air's moiſture to be perceived. For I ſuppos'd, that after a time this unuſual ſtretch of the Rope would ceaſe, and when the weight as ſuch could not lengthen it any more, it would then be capable of being contracted or relax'd, according as the weather ſhould be moiſt or dry, and ſo afford me a kind of Hygroſcope. Upon theſe grounds I firſt caus'd a Rope that was about 20 or 22 yards in length, but of no great thickneſs, to have one of its ends faſten'd to an immoveable Body at a convenient height from the ground, and then caus'd a Pully to be ſo faſten'd to another ſtable Body at the di
ſtance

ftance of 18 or 20 yards from the firft; that the
Rope, refting upon the Pully, lay almoft horizontally. But to the end of that part of the Rope,
wh ch from the Pully reach'd within two or three
foot of the ground, was faften'd by a Ring a Leaden
weight of at leaft fifty pound. To which was alfo
faften'd a light *Index* plac'd horizontally, whofe end
moved along an erected board, which by tranfverfe
lines was divided into inches and parts of inches,
reaching both a good way upwards and downwards,
that the *Index* might within thofe bounds have room
to play up and down according to the alterations of
the weather.

It being then Summer, this Tryal was made in a
Garden, though partly under a Penthoufe, that the
Rope might be more expos'd to the Air than it
would have been within doors; and two or three
dayes, if I mifremember not the time, were fpent,
before the weight had brought the rope to the utmoft ftretch it was able to give it, after which it began mani.eftly to fhrink and lengthen according to
the weather. And I find in one of my Notes, that
once I look'd, when I was ready to go to bed, upon
the fufpended weight, and mark'd how low it reach'd
upon the divided board ; and that a great part of the
night having been rainy, looking again about half an
hour after eight in the morning, I found the Cord fo
fhrunk, that the weight was rais'd above five inches,
and yet the day growing dry and windy, and fometimes warm, the weight had at night ftretched the
Rope more than the moifture had contracted it the
day before.

Afterwards having procur'd a far greater weight,
but therefore unapt to be near fo much rais'd, I fub-
ftituted

ſtituted it in the place of that formerly mention'd, and having ſuffer'd it to ſtretch the rope as far as it could, I made and regiſter'd ſome Obſervations, two whereof having been preſerved, I ſhall tranſcribe them juſt as I find them.

June the 4th. At half an hour after nine of the clock at night, I look'd upon the hundred pound weight that hung at the bottom of the rope, the weather being then fair, and a mark being put at that part of the erected board, where the bottom of the weight touched; I perceiv'd the sky a while after to grow cloudy and overcaſt, but without rain; wherefore going to view the weight again, I found it to be riſen a quarter of an inch or more, and, looking on my Watch, perceiv'd there had paſſed an hour and quarter ſince the mark was made.

June the 6th. Being not well yeſterday, the weight was obſerv'd by two of my ſervants, and it then reſted at the eleventh inch of the erected board. This morning about eight of clock I viſited it my ſelf, and found it to be riſen about half a quarter of an inch above the eighth inch, the morning being cloudy, though the ground very dry and duſty. The weather being more overcaſt, within ſomewhat leſs than an hour afterwards I viſited the weight again, (ſome ſcatter'd drops of rain then beginning to fall,) and found it to be riſen about half an inch above the newly mention'd eighth mark. How much more the rope would have been contracted in ſuch laſting moiſt weather, as uſually happens in Winter, I cannot ſay, having been reduc'd to break off the Experiment, upon a removal, I was, long before that ſeaſon, oblig'd to make.

I am forry I cannot add my other obfervations, but thefe I hope may fuffice to let you fee, that the force of the Air's moifture is not fmall, fince it could raife fuch a weight as an hundred pound, efpecially confidering the flendernefs of the rope it affected. For having meafur'd the Diameter near the weight, I found it (as one of my Notes informs me) to be but about the third part of an Inch.

'Twas $\frac{3}{10}$ and 4 decimal parts of $\frac{1}{10}$.

F I N I S.

www.ingramcontent.com/pod-product-compliance
Lightning Source LLC
Chambersburg PA
CBHW021710210326
41599CB00013B/1600